Biochemical and Clinical Aspects of Neuropeptides

Synthesis, Processing, and Gene Structure

Academic Press Rapid Manuscript Reproduction

Proceedings of the 5th Conference on Macromolecular Synthesis
Held in Hamburg–Blankenese, Federal Republic of Germany,
May 29–June 2, 1983

Biochemical and Clinical Aspects of Neuropeptides

Synthesis, Processing, and Gene Structure

Edited by

Gebhard Koch

Physiologisch–Chemisches Institut
Abteilung Molekularbiologie
Universität Hamburg
Hamburg, Federal Republic of Germany

Dietmar Richter

Physiologisch–Chemisches Institut
Abteilung Zellbiochemie
Universität Hamburg
Hamburg, Federal Republic of Germany

1983

ACADEMIC PRESS, INC.
(Harcourt Brace Jovanovich, Publishers)
Orlando San Diego San Francisco New York London
Toronto Montreal Sydney Tokyo São Paulo

ACADEMIC PRESS, INC.
Orlando, Florida 32887

United Kingdom Edition published by
ACADEMIC PRESS, INC. (LONDON) LTD.
24/28 Oval Road, London NW1 7DX

Conference on Macromolecular Synthesis (5th : 1983 : Blankenese, Hamburg, Germany)
Biochemical and clinical aspects of neuropeptides.

Includes index.
1. Neuropeptides--Congresses. I. Koch, Gebhard.
II. Richter, Dietmar, Date . III. Title.
QP552.N39C66 1983 612'.814 83-22381
ISBN 0-12-417320-9 (alk. paper)

PRINTED IN THE UNITED STATES OF AMERICA

83 84 85 86 9 8 7 6 5 4 3 2 1

Contents

PART II

Novel Neuropeptides

PART III

Neuropeptide Deficiencies

PART IV

Posttranslational Processing

Contributors

Numbers in parentheses indicate the pages on which the authors' contributions begin.

Felicia B. Axelrod (113), *Department of Pediatrics, New York University Medical Center, New York, New York*

Karl Bauer (195), *Institut für Biochemie, Technische Universität Berlin, Berlin, Federal Republic of Germany*

Graeme I. Bell (293), *Department of Biochemistry and Biophysics, University of California, San Francisco, California*

Heinz Bodenmüller (75), *Department of Biophysics, Max-Planck-Institut for Medical Research, Heidelberg, Federal Republic of Germany*

Xandra O. Breakefield (113), *Department of Human Genetics, Yale University School of Medicine, New Haven, Connecticut*

J. Peter H. Burbach (211), *Rudolf Magnus Institute for Pharmacology, Medical Faculty, University of Utrecht, Utrecht, the Netherlands*

Raymond Carroll (3), *Department of Biochemistry, The University of Chicago, Chicago, Illinois*

Carmela M. Castiglione (113), *Department of Human Genetics, Yale University School of Medicine, New Haven, Connecticut*

Shu Jin Chan (3), *Department of Biochemistry, The University of Chicago, Chicago, Illinois*

Harry M. Charlton (129), *Department of Anatomy, Oxford University, Oxford, England*

Michel Chrétien (83), *Protein and Pituitary Hormone Laboratory, Clinical Research Institute of Montreal, Montreal, Canada*

Olivier Civelli (45), *Department of Chemistry, University of Oregon, Eugene, Oregon*

Michael Comb (45), *Department of Chemistry, University of Oregon, Eugene, Oregon*

Lisa Coussens (113), *Genentech, Inc., South San Francisco, California*

Robert J. Crawford (293), *Department of Biochemistry and Biophysics, University of California, San Francisco, California*

David de Wied (211), *Rudolf Magnus Institute for Pharmacology, Medical Faculty, University of Utrecht, Utrecht, the Netherlands*

Kevin Docherty (3), *Department of Biochemistry, The University of Chicago, Chicago, Illinois*

Graham J. Dockray (161), *Physiological Laboratory, University of Liverpool, Liverpool, England*

James Douglass (45), *Department of Chemistry, University of Oregon, Eugene, Oregon*

Thomas J. Dull (277), *Genentech, Inc., South San Francisco, California*

Ronald M. Evans (267), *Tumor Virology Laboratory, The Salk Institute, La Jolla, California*

Javier Garzón (59), *Department of Pharmacology, University of California, San Francisco, California*

Jean-Pierre Gautron (235), *Unité de Neuroendocrinologie, INSERM, Paris, France*

Marie J. Gibson (129), *Division of Endocrinology, Mount Sinai School of Medicine, New York, New York*

Andrew Giraud (161), *Physiological Laboratory, University of Liverpool, Liverpool, England*

Alane Gray (277), *Genentech, Inc., South San Francisco, California*

Edward Herbert (45), *Department of Chemistry, University of Oregon, Eugene, Oregon*

Peter Hobart (293), *Department of Biochemistry and Biophysics, University of California, San Francisco, California*

Volker Höllt (59), *Max-Planck-Institut für Psychiatrie, Munich, Federal Republic of Germany*

Werner Hoffmann (175), *Institute for Molecular Biology, Austrian Academy of Sciences, Salzburg, Austria*

Bernhard Horsthemke (195), *Institute für Biochemie, Technische Universität Berlin, Berlin, Federal Republic of Germany*

Kuo-Liang Hsi (83), *Protein and Pituitary Hormone Laboratory, Clinical Research Institute of Montreal, Montreal, Canada*

Anton Hutticher (175), *Institute for Molecular Biology, Austrian Academy of Sciences, Salzburg, Austria*

Richard Ivell (245, 259), *Physiologisch–Chemisches Institut, Abteilung Zellbiochemie, Universität Hamburg, Hamburg, Federal Republic of Germany*

Claude Kordon (235), *Unité de Neuroendocrinologie, INSERM, Paris, France*

Gabor L. Kovács (211), *Rudolf Magnus Institute for Pharmacology, Medical Faculty, University of Utrecht, Utrecht, the Netherlands*

Günther Kreil (175), *Institute for Molecular Biology, Austrian Academy of Sciences, Salzburg, Austria*

Dorothy T. Krieger (129), *Division of Endocrinology, Mount Sinai School of Medicine, New York, New York*

Choh Hao Li (17), *Hormone Research Laboratory, University of California, San Francisco, California*

Horace H. Loh (59), *Department of Pharmacology, University of California, San Francisco, California*

Gerard Martens (45), *Department of Chemistry, University of Oregon, Eugene, Oregon*

Fred Nyberg (99), *Department of Pharmacology, University of Uppsala, Uppsala, Sweden*

Edward Oates (45), *Department of Chemistry, University of Oregon, Eugene, Oregon*

Gregory Orloff (113), *Department of Human Genetics, Yale University School of Medicine, New Haven, Connecticut*

Eliane Pattou (235), *Unité de Neuroendocrinologie, INSERM, Paris, France*

Mark J. Perlow (129), *Department of Neurology, University of Illinois School of Medicine, West Side Veterans Administration Hospital, Chicago, Illinois*

Joe R. Reeve, Jr. (161), *CURE, Veterans Administration Hospital Center, Los Angeles, California*

Jens F. Rehfeld (151), *University Department of Clinical Chemistry, Rigshospitalet, Copenhagen, Denmark*

Dietmar Richter (185, 245, 259), *Physiologisch–Chemisches Institut, Abteilung Zellbiochemie, Universität Hamburg, Hamburg, Federal Republic of Germany*

Klaus Richter (175), *Institute for Molecular Biology, Austrian Academy of Sciences, Salzburg, Austria*

Jean Rivier (225), *Peptide Biology Laboratory, The Salk Institute, La Jolla, California*

Haim Rosen (45), *Department of Chemistry, University of Oregon, Eugene, Oregon*

Michael G. Rosenfeld (267), *School of Medicine, University of California, San Diego, La Jolla, California 92093*

William J. Rutter (293), *Department of Biochemistry and Biophysics, University of California, San Francisco, California*

Ray Sanchez-Pescador (293), *Department of Biochemistry and Biophysics, University of California, San Francisco, California*

Blanca San Segundo (3), *Department of Biochemistry, The University of Chicago, Chicago, Illinois*

H. Chica Schaller (75), *Department of Biophysics, Max-Planck-Institut for Medical Research, Heidelberg, Federal Republic of Germany*

Hartwig Schmale (185), *Physiologisch–Chemisches Institut, Abteilung Zellbiochemie, Universität Hamburg, Hamburg, Federal Republic of Germany*

James Scott (293), *Department of Biochemistry and Biophysics, University of California, San Francisco, California*

Nabil G. Seidah (83), *Protein and Pituitary Hormone Laboratory, Clinical Research Institute of Montreal, Montreal, Canada*

Bernd R. Seizinger (59), *Max-Planck-Institut für Psychiatrie, Munich, Federal Republic of Germany*

Mark Selby (293), *Department of Biochemistry and Biophysics, University of California, San Francisco, California*

Lu-ping Shen (293), *Department of Biochemistry and Biophysics, University of California, San Francisco, California*

Joachim Spiess (225), *Peptide Biology Laboratory, The Salk Institute, La Jolla, California*

Donald F. Steiner (3), *Department of Biochemistry, The University of Chicago, Chicago, Illinois*

Lars Terenius (99), *Department of Pharmacology, University of Uppsala, Uppsala, Sweden*

Sidney Udenfriend (33), *Roche Institute of Molecular Biology, Roche Research Center, Nutley, New Jersey*

Axel Ullrich (113, 277), *Genentech, Inc., South San Francisco, California*

Wylie Vale (225), *Peptide Biology Laboratory, The Salk Institute, La Jolla, California*

Xin-Chang Wang (211), *Rudolf Magnus Institute for Pharmacology, Medical Faculty, University of Utrecht, Utrecht, the Netherlands*

Earl A. Zimmerman (129), *Department of Neurology, Columbia Presbyterian Hospital, New York, New York*

Preface

Research on neuropeptides was initiated two decades ago with the discovery of small peptides that have profound effects on certain biological events. In the past few years, due to the efforts of several laboratories, great progress has been achieved in this field. Thus we chose "Biochemical and Clinical Aspects of Neuropeptides" as the topic for our yearly conference on Macromolecular Synthesis. Clinical aspects were included to fill the needs of the basic researcher whose interests lie in the application of research findings.

In our meeting three years ago on "Biosynthesis, Modification, and Processing of Cellular and Viral Polyproteins," the focus was on the synthesis and processing of polyproteins. It became apparent then that many neuropeptides are synthesized in the form of larger precursor polyproteins, sometimes exceeding ten times the size of the known functional peptides. Like other proteins that leave the cell, the precursors are synthesized on membranes of the rough endoplasmic reticulum and are then processed and packaged into secretory vesicles.

At this year's meeting it became evident that the number of identified polyproteins had increased considerably. Many of the polyproteins may serve as precursors to several functionally distinct biological entities due to a series of specific proteolytic cleavages. In spite of intensive research we still know little about the mechanism and regulation of the processing and packaging events.

Another feature that excites much speculation is the finding that many of the peptides present in the mammalian brain are now turning up in a variety of other animals including invertebrates. Neuropeptides are also being discovered in many peripheral organs such as the intestine, thyroid, and gonads, and peptides thought to be restricted to peripheral organs turn up in the central nervous system. This suggests a much wider diversification of peptide functions than could have been imagined a few years ago.

Recent advances in DNA cloning techniques have led to the elucidation of the structural organization of corresponding genes for neuropeptides. Based on sequence analysis many new potential peptides can be predicted. Research is now needed to determine the physiological functions of these peptides.

Considerable research effort is required to solve the many exciting questions which presently remain unanswered. This is especially true for clinical aspects and applications. Since peptides do not naturally cross the blood–brain barrier their pharmacological analysis is hindered. One of the immediate goals will be to study the triggering mechanism and the control of the release of neuropeptides. We certainly can expect that in another three years our knowledge in this field will have increased substantially.

We are grateful to the Stiftung Volkswagenwerk, the Behörde für Wissenschaft und Forschung, Hansestadt Hamburg, and the Hamburgische Wissenschaftliche Stiftung for financial support.

Introduction

INTRACELLULAR PROTEOLYTIC MECHANISMS IN THE BIOSYNTHESIS OF HORMONES AND PEPTIDE NEUROTRANSMITTERS*

Donald F. Steiner
Kevin Docherty
Shu Jin Chan
Blanca San Segundo
Raymond Carroll

Department of Biochemistry
The University of Chicago
Chicago, IL

I. INTRODUCTION

Many small bioactive peptides and peptide neurotransmitters are derived biosynthetically from larger precursor forms which are processed intracellularly during their synthesis and migration through the cell into storage or transport vesicles. Historically, the first of these proteins to be isolated and studied was proinsulin, the single chain precursor of insulin which contains the B and A chains linked through a 35 residue connecting polypeptide segment (49,51). On the basis of their early studies Sachs and coworkers had suggested that vasopressin and oxytocin, the small nonapeptide hormones of the posterior pituitary and supraoptic nucleus, might be derived from a peptide whose synthesis was inhibited by puromycin and which differed in its properties from mature vasopressin (for a review see ref. 44). The macromolecular nature of this putative precursor has only recently been established with certainty (32). The discovery of proinsulin in 1967 placed this area on a solid footing (47) and was followed after about five years by the discovery of other precursors, first of parathyroid hormone (38) and later of many other peptide hormones and neurotransmitters as well as of some cellular and viral proteins (18,27). As new neuropeptides continue to be discovered the structures of their precursors are rapidly being elucidated by means of gene cloning technology yielding

**Work from this laboratory was supported by USPHS Grants #AM13914 and AM20595.*

ISBN 0-12-417320-9

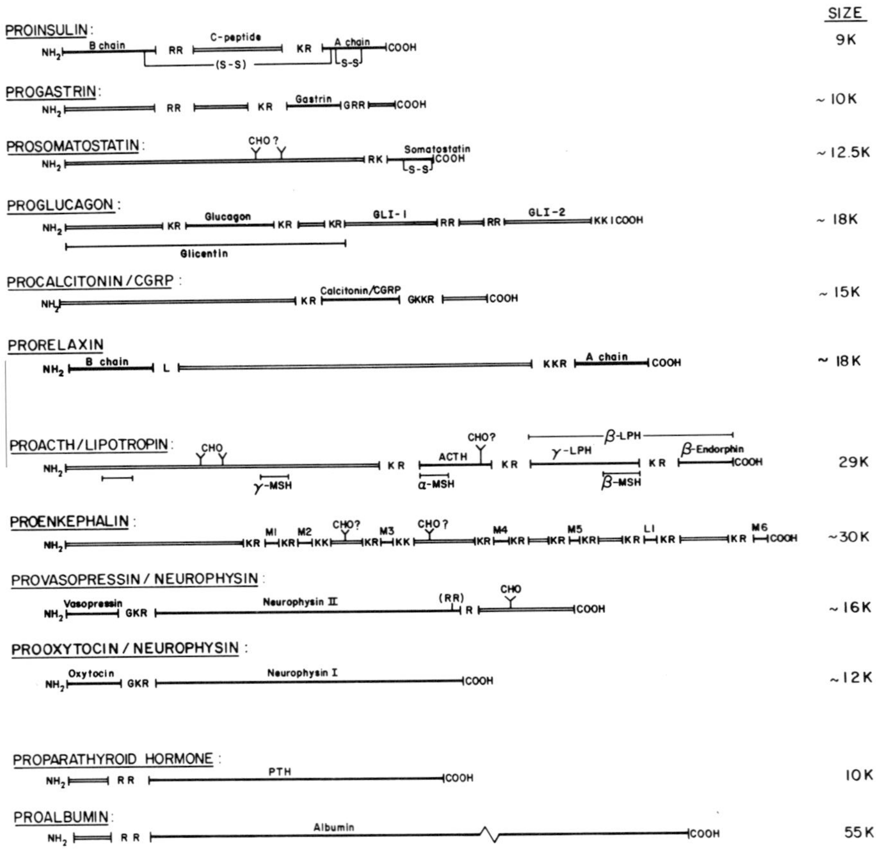

FIGURE 1: Schematic structures of some of the known proproteins showing residues at cleavage sites. Heavy single lines indicate regions which appear as biologically active products. Data sources are as follows: proinsulin (52), progastrin (56), prosomatostatin (52), proglucagon (4), procalcitonin/CGRP (2), prorelaxin (20), proACTH/lipotropin (33,52), proenkephalin (10), provasopressin/neurophysin (32), prooxytocin/neurophysin (39), proparathyroid hormone (38), proalbumin (52).

a wealth of new information to be digested and correlated with their biosynthesis and physiology. This symposium serves admirably to underscore the remarkable progress in this field due to these new methodologies.

The structures of some of the known prohormones are summarized schematically in Figure 1. It is evident that most of these are as large or larger than proinsulin, and that the biologically active regions within them are dispersed along their polypeptide chains in what appears to be a random pattern, usually making up only a very small part of the total precursor protein. The active regions in some precursors are tandemly repeated and in some cases more than one kind of active region may be contained within a single polypeptide, as illustrated especially by the ACTH (33) and enkephalin (10) precursors, somewhat in analogy with the polyprotein precursors of many viruses. However, it is not yet clear whether these proteins consist of more than one complete folded domain, defined as a unit of organized secondary and tertiary structure, as is usually the case with the much larger viral polyproteins (27). To date no 3-dimensional studies on precursor proteins have been reported.

The processing of these precursors is currently a subject of great interest since these intracellular proteolytic mechanisms clearly can regulate the nature and properties of the bioactive peptides that are derived from any given precursor and released into the circulation or within the nervous system at selected local sites and synaptic junctions. The first step of processing in almost all cases represents the cleavage of the signal peptide from the N-terminally extended presecretory form of the precursor that is the initial translational product on the ribosomes (5,26,34). The signal peptidase is localized in the membrane of the rough endoplasmic reticulum (22) and operates very rapidly, probably mainly cotranslationally (28,52) to remove the N-terminal extension. This process is essentially universal to all secreted proteins.

In this brief review we will focus in greater detail on the proteolytic mechanisms involved in proprotein processing, their intracellular localization, and their role(s) in generating a diverse assortment of secreted products in a growing variety of neuroendocrine cells. It is especially useful to consider the possible common features in these processing mechanisms which may allow a rather limited repertoire of cellular proteases to accomplish the processing of many different precursors in either the same or different neuroendocrine cells.

II. PROPROTEIN PROCESSING

In contrast to prepeptide processing which occurs very rapidly during synthesis or shortly after peptide chain completion, proprotein processing begins only 10-20 minutes after synthesis and folding of the peptide chain has been accomplished in the RER and the peptides have been transferred to the Golgi apparatus where concentration and secretion granule formation begins (23,48). The movement from RER to the Golgi is an energy requiring process which involves the budding of small transport vesicles from smooth regions of the RER and their fusion with the cis elements of the Golgi apparatus (14,23,50). Biochemical and radioautographic studies (14,19,37,50) have confirmed that processing begins within the Golgi apparatus for a number of different precursors and often continues for many hours as these products continue to undergo maturation within newly-formed secretion granules which bud off from the trans Golgi elements (17,25). Thus isolated islet secretion granules containing biosynthetically labeled proinsulin and other islet prohormones continue to process these materials for many hours when incubated under appropriate conditions in vitro (25). As conversion is completed the insulin storage granules undergo further morphological changes with condensation of the insulin and other precursor-derived materials, including the C-peptide, which are then all released coordinately from the beta cell during secretion (42,51).

As indicated in Fig. 2 this sequence of cellular events also occurs within the neurosecretory cell and the long period of axonal

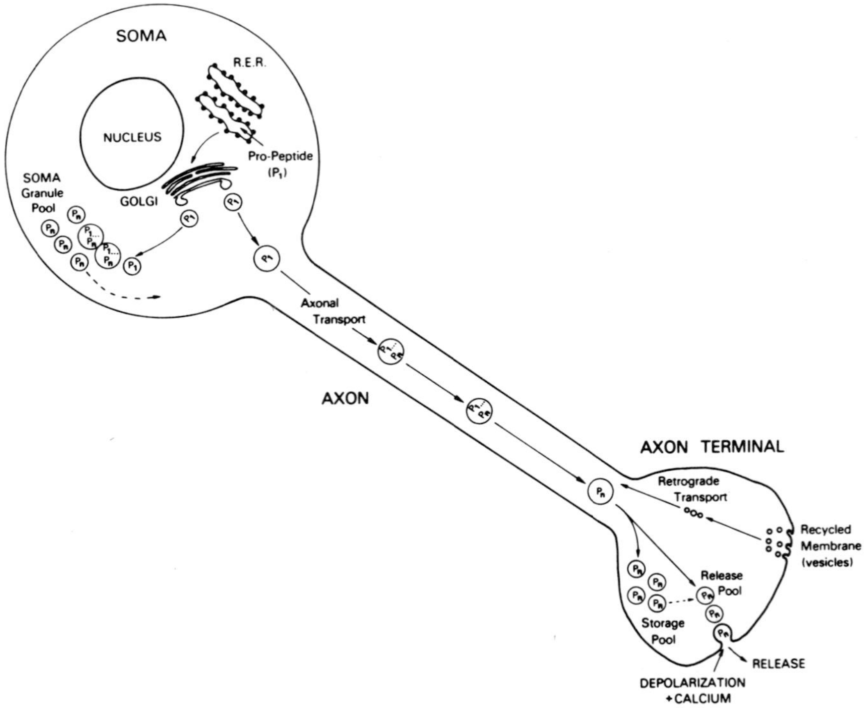

FIGURE 2. Schematic diagram illustrating the biosynthesis and intracellular transport of a typical neurosecretory cell. (Reproduced with permission of the authors from ref. 16).

transport may provide ample time for the maturation of products (16). In the case of proinsulin, if one assumes that the processing enzymes are as active as e.g. pancreatic trypsin then the amounts required to achieve the observed in vivo rates of conversion within the highly concentrated secretion granule are quite small, amounting to less than 0.1% of the amount of precursor. It is this fact, coupled with difficulties of efficiently isolating and recovering intact Golgi and secretion granule fractions free of other cellular organelles (such as lysosomes) that has complicated the identification of the converting protease(s) in most tissues.

III. NATURE OF THE PROCESSING PROTEASES

Inspection of the data in Figure 1 indicates that the most frequent sites of cleavage in all precursor proteins is at pairs of basic amino acids such as Arg-Arg or Lys-Arg. This highly characteristic feature has led to the suggestion that trypsin-like and carboxypeptidase B-like enzymes working in concert could generate the correct natural products from many of these precursors, and work with model systems has verified this in the case of proinsulin (24,51). Abnormal proinsulins and proalbumins have been described in which mutations have occurred in the basic residue pairs at cleavage sites leading to the substitution of a neutral amino acid for one of the basic residues, i.e. Arg to a neutral residue in an abnormal proinsulin (40), Arg to Gln in the Christchurch proalbumin (6) and Arg to His in proalbumin Lille (1). In both cases normal processing is prevented or greatly retarded by this change. Replacement of the Arg or Lys residues of anglerfish islet proinsulin, proglucagon and prosomatostatin by the analogs conavanine and thialysine during biosynthesis prevented cleavage of these prohormones (35), providing further evidence for the essential role in this process of dibasic amino acids.

That cleavage does not always require pairs of basic amino acids is indicated by chicken proalbumin which in contrast to the human and rat precursors has Ala-Arg at the C-terminus of the propeptide and is readily processed (41). Several other precursors where cleavages at a single basic amino acid (usually arginine) or at other special sites have been recognized include dog proinsulin (31) which contains a processed single arginine site in its C-peptide (position C8), the vasopressin-neurophysin precursor (see Fig. 1) and the precursor of canine pancreatic polypeptide (46). Several single basic residue processing sites are also known to occur in CCK (43). It remains unclear whether all of these sites are processed by the same trypsin-like enzyme(s) which recognize the dibasic pairs or may be acted on by another class of trypsin-like proteases during transit from the ER to the storage granules. In the case of canine pancreatic polypeptide, biosynthetic studies on cultured dog F cells support the idea that a different protease cleaves these sites. Thus, after prolonged culture the ability to cleave at the single basic site is lost while the ability to cleave at other sites is retained (46).

Another less frequently used processing site involves cleavage between Leu-Ala or Leu-Ser residues, an event requiring an enzyme having proteolytic specificity more akin to that of chymotrypsin than of trypsin. Such a sequence is cleaved in the human, pork and rat proinsulin C-peptides, but usually only 20-30% of the C-peptide is processed at this position and it has accordingly been viewed as an incidental or minor cleavage (8,30,53). However, the recent finding of Hudson et al (20) that the relaxin precursor, a homologue of proinsulin (see Fig. 1), is probably processed mainly at a Leu-Ser site on the N-terminal side of its very long connecting peptide segment, while the cleavage site between the connecting segment and the A chain is made up of 3 basic amino acids, suggests that chymotrypsin-like cleavage may be a major processing mechanism in some systems. All these variations taken together suggest that a number of different enzymic activities may normally participate in prohormone processing and that the selection of cleavage sites may be dictated in part by the nature and relative amounts of these enzymes as well as by the primary, secondary and tertiary structure of the precursor in the region(s) to be processed, particularly their accessibility and susceptibility to proteolysis by subsets of locally available proteases.

A variety of serine proteases have been proposed as prohormone cleavage enzymes, including kallikreins (36) and plasmin or plasminogen activator (55). Although some of these enzymes may be involved in cleavages at single basic residues, presently available evidence in several processing systems, including islets of Langerhans and pituitary, has implicated thiol proteases having properties similar to cathepsin B as the main processing enzymes (9,12,15). We have recently identified several thiol-containing proteases in partially purified rat islet granule fractions, including one of 31.5 kilodaltons which is apparently identical with cathepsin B, and a somewhat larger thiol protease with an estimated molecular size of about 39 kilodaltons, using a sensitive site-specific labeled probe, ^{125}I-Tyr-Ala-Lys-Arg chloromethyl ketone. The 39 kilodalton component also appears to be related to cathepsin B since treatment with pepsin converts it to a form which is similar in size to cathepsin B and is immunoprecipitated by an antiserum to cathepsin B (Docherty, K. unpublished data). In further studies using more highly purified insulin secretory granules from an islet cell tumor (21) it has been possible to show that both the higher and lower mol. wt. forms of this cathepsin B-like material are present in both lysosomes and secretion granules while cathepsin H appears only in the lysosomes. Moreover, both of these components label more readily with the dibasic chloromethylketone than with a comparable probe containing only a single arginine residue (Docherty, K. unpublished results).

In contrast to the results obtained with islet secretion granules, studies in our laboratory on secretion granules purified from calf adrenal medullary tissue, a probable site of processing of proenkephalins, did not reveal the presence of significant amounts of these cathepsin B-related components, although they were clearly present in the lysosomes of this tissue. This last finding along with our experience that purified cathepsin B (from rat liver lysosomes) does not correctly cleave proinsulin and in fact degrades it at high concentrations casts some doubt on the possibility of a general role of cathepsin B per se in prohormone conversion. Some workers have reported conversion of proinsulin by cathepsin B (3) but the cleavage products were not thoroughly characterized. Various additional lines of evidence indicating that ordinary lysosomal cathepsin B is unlikely to be the converting protease for proinsulin or other prohormones has been summarized elsewhere (11,13).

Recent studies in our laboratory indicate that cathepsin B is derived in biosynthesis from a larger precursor of approximately 44 kilodaltons (Steiner, D. and Carroll, R., unpublished data). A faint band of this size is also labeled by ^{125}I-Tyr-Ala-Lys-Arg chloromethylketone in the islet secretion granule fractions. These findings raise the interesting possibility that larger or modified forms of cathepsin B derived from or related to this precursor may be cosegregated into secretion granules along with proinsulin and participate in its conversion, while the bulk of cathepsin B may be directed to the lysosomes.

Whether essentially the same converting enzymes act in all tissues, or individually specialized proteases exist in each endocrine or neural cell type engaged in the processing of precursors remains an important issue to be settled. If multiple enzymes are involved which are derived in evolution from the catheptic thiol proteases then this phylogenetically ancient superfamily of proteases, which also includes plant enzymes such as papain (54), could be expected to include dozens of related gene products corresponding to individually specialized tissue-specific processing enzymes. Studies on the genes for these enzymes which have been initiated in our laboratory may ultimately provide the answer to this question.

It is now clear that proteolytic processing is an ancient and widespread mechanism for producing small biologically active peptides. Processing at pairs of basic residues has been found to occur in primitive vertebrates (7) as well as in invertebrates (45) and recently also even in yeast (29). The yeast α-mating factor, a small secreted peptide, is derived in biosynthesis from a larger precursor protein which contains a long nonrepetitive N-terminal segment followed by 4 tandemly arranged copies of a peptide containing the α-mating factor sequence separated by pairs of basic residues which serve as processing sites (29). The nature of the yeast processing enzyme also is not yet known, but is obviously of considerable interest.

IV. SUMMARY AND CONCLUSIONS

1. Precursor processing is a phylogenetically ancient and widely used mechanism for producing biologically active peptides. The use of N-terminal signal peptides to accomplish secretion seems to be a universally occurring mechanism, while proprotein processing may be a more prominent feature of eukaryotic cells.
2. Maturation of proproteins begins only after transport to the Golgi apparatus has been completed and probably continues for many hours within newly formed secretory vesicles as these are stored in the cytosol or transported along axons to more peripheral sites of release.
3. The requirement for paired basic residues for efficient proteolysis in most precursors suggests that a single (or only a few) trypsin-like protease(s) has evolved which cleaves with great site selectivity but retains the ability to process many sites within the same precursor or in different precursors within the same cell, or in different cells or tissues. (The restriction endonucleases, which selectively cleave DNA molecules only at certain defined nucleotide sequences, provide a reasonable analogy of this kind of "highly selective non-specificity" in processing.)
4. Cleavage is probably strongly influenced by other factors such as secondary and tertiary structure, but at present there is no definitive information on the 3-dimensional structure of any precursor protein. Other modifications (glycosylation, phosphorylation, sulfation) are not known to influence processing patterns.
5. Golgi/granule processing may also rarely occur at sites other than pairs of basic amino acids, including single basic residues (trypsin-like), Leu-Ala, Leu-Ser, or Tyr-Ala bonds (chymotrypsin-like) as well as other specialized non-tryptic cleavages, suggesting that a mixture of proteases occurs in the Golgi/granule system and may cleave susceptible sites in some precursors.
6. Because of the dual role of the Golgi apparatus in sorting secreted products from lysosomal hydrolases, an intimate relationship may exist between certain of the lysosomal proteases, e.g. especially cathepsin B or its precursors, and the Golgi/granule processing systems.
7. The sequestration of the proteolytic products of precursors within secretion granules often leads to the coordinate discharge of highly complex mixtures of peptides having related or overlapping biological activities, as well as of purely non-functional peptide "left-overs"

(e.g. the proinsulin C-peptide) which can serve as useful markers of secretion or localization (42), as well as of evolutionary relationships.

8. Errors in cleavage due to point mutations in precursors have been identified in the case of proinsulin and proalbumin leading to the accumulation of incorrectly processed materials. These and/or defects in converting proteases per se may ultimately be found to contribute to the pathogenesis of some neuro-endocrine disorders.

ACKNOWLEDGMENTS

We are indebted to Dr. H. Kirschke (Halle, GDR) for generously supplying an antiserum to cathepsin B and to Ms. Lisa Fuller for expert assistance in preparing this manuscript.

REFERENCES

1. Abdo, Y., Rousseaux, J., and Dautrevaux, M., FEBS Lett. 131, 286 (1981).
2. Amara, S. G., Jonas, V., Rosenfeld, M. G., Ong, E. S., and Evans, R. M., Nature 298, 240 (1982).
3. Bansal, R., Ahmad, N., and Kidwai, J. R., Acta Diabetica Latina 17, 255 (1980).
4. Bell, G. I., Santerre, R. F., and Mullenbach, G. T., Nature 302, 716 (1983).
5. Blobel, G., and Dobberstein, B., J. Cell. Biol. 67, 835 (1975).
6. Brennan, S.O., and Carrell, Nature 274, 908 (1978).
7. Chan, S. J., Emdin, S. O., Kwok, S. C. M., Kramer, J. M., Falkmer, S., and Steiner, D. F., J. Biol. Chem. 256, 7595 (1981).
8. Chance, R. E., in "Proceedings of the 7th Congress of the International Diabetes Federation" (R. R. Rodriquez, and J. J. Vallance-owen, eds.), p. 292 Excerpta Medica, (1971).
9. Chang, T-L., and Loh, Y. P., Endocrinology 112, 1832 (1983).
10. Comb, M., Seeburg, P. H., Adelman, J., Eiden, L., and Herbert, E., Nature 295, 663 (1982).
11. Docherty, K., and Steiner, D. F., Ann. Rev. Physiol. 44, 625 (1981).
12. Docherty, K., Carroll, R. J., and Steiner, D. F., Proc. Natl. Acad. Sci. USA 79, 4613 (1982).
13. Docherty, K., Carroll, R., and Steiner, D. F., Proc. Natl. Acad. Sci. USA 80, 3245 (1983).
14. Farquhar, M. G., and Palade, G. E., J. Cell. Biol. 91, 77S (1981).

15. Fletcher, D. J., Quigley, J. P., Bauer, G. E., and Noe, B. D., J. Cell. Biol. 90, 312 (1981).
16. Gainer, H., Sarne, Y., and Brownstein, M. J., J. Cell. Biol. 73, 366 (1977).
17. Habener, J. F., and Potts, J. T. Jr., Endocrinology 104, 265 (1979).
18. Holzer, H., and Heinrich, P. C., Ann. Rev. Biochem. 49, 63 (1980).
19. Howell, S. L., Nature New Biology 235, 85 (1972).
20. Hudson, P., Haley, J., Cronk, M., Shine, J., and Niall, H., Nature 291, 127 (1981).
21. Hutton, J. C., Penn, E. J. and Peshavaria, M., Diabetologia 23, 365 (1982).
22. Jackson, R. C., and Blobel, G., Proc. Natl. Acad. Sci. USA 74, 5598 (1977).
23. Jamieson, J. D., and Palade, G. E., in "International Cell Biology" (B. R. Brinkley, and H. R. Poles, eds.) p. 308. Rockefeller University Press, New York (1977).
24. Kemmler, W., Peterson, J. D., and Steiner, D. F., J. Biol. Chem. 246, 6786 (1971).
25. Kemmler, W., Steiner, D. F., and Borg, J., J. Biol. Chem. 248, 4544 (1973).
26. Kemper, B., Habener, J. F., Mulligan, R. C., Potts, J. T. Jr., and Rich, A., Proc. Natl. Acad. Sci. USA 71, 3431 (1974).
27. Korant, B., in "Molecular Biology of Picornaviruses" (R. Perez-Bucoff, ed.) p. 149. Plenum Publishers, New York (1979).
28. Kreil, G., Ann. Revs. Biochem. 50, 317 (1981).
29. Kurjan, J., and Herskowitz, I., Cell 30, 933 (1982).
30. Kuzuya, H., Chance, R. E., Steiner, D. F., and Rubenstein, A. H., Diabetes 27, 161 (1978).
31. Kwok, S. C. M., Chan, S. J., and Steiner, D. F., J. Biol. Chem. 258, 2357 (1983).
32. Land, H., Schütz, G., Schmale, H., and Richter, D., Nature 295, 299 (1982).
33. Mains, R. E., Eipper, B. A., and Ling, N., Proc. Natl. Acad. Sci. USA 74, 3014 (1977).
34. Milstein, C., Brownlee, G. G., Harrison, T. M., and Mathews, M. B., Nature New Biol. 239, 117 (1972).
35. Noe, B. D., J. Biol. Chem. 256, 4940 (1981).
36. Ole-Moi Yoi, O., Seldin, D. C., Spragg, J., Pinkus, G. S., and Austen, K. F., Proc. Natl. Acad. Sci. USA 76, 3612 (1979).
37. Orci, L., Lambert, A. E., Kanazawa, Y., Amherdt, M., Rouiller, C., and Renold, A. E., J. Cell. Biol. 50, 565 (1971).
38. Potts, J. T. Jr., Kronenberg, H. M., and Rosenblatt, M., Adv. Prot. Chem. 35, 323 (1982).
39. Richter, D., and Schmale, H., this volume.
40. Robbins, D. C., Blix, P. M., Rubenstein, A. H., Kanazawa, Y., Kosaka, K., and Tager, H. S., Nature 291, 679 (1981).

41. Rosen, A. M., and Geller, D. M., Biochem. Biophys. Res. Commun. 78, 1060 (1977).
42. Rubenstein, A. H., Steiner, D. F., Horwitz, D. L., Mako, M. E., Block, M. B., Starr, J. L., Kuzuya, H., and Melani, F., Recent Progress in Hormone Research 33, 435 (1977).
43. Ryder, S. W., Straus, E., and Yalow, R. S., Proc. Natl. Acad. Sci. USA 77, 3669 (1980).
44. Sachs, H., Fawcett, P., Takabatake, Y. and Portonova, R., Rec. Prog. Horm. Res. 25, 447 (1969).
45. Schueller, R. H., Jackson, J. F., McAllister, L., Schwartz, J., Kandel, E. R. and Axel, R., Cell 28, 707 (1982).
46. Schwartz, T. W. and Tager, H. S., Hoppe Seyler's Z. Physiol. Chem. 363, 889 (1982).
47. Steiner, D. F., and Oyer, P. C., Proc. Natl. Acad. Sci. USA 57, 473 (1967).
48. Steiner, D. F., Trans. N.Y. Acad. Sci., Ser. II 30, 60 (1967).
49. Steiner, D. F., Clark, J. L., Nolan, C., Rubenstein, A. H., Margoliash, E., Aten, B. and Oyer, P.E., Rec. Prog. Horm. Res. 25, 207 (1969).
50. Steiner, D. F., Clark, J. L., Nolan, C., Rubenstein, A. H., Margoliash, E., Melani, F., and Oyer, P. E., in "Nobel Symposium 13, on the Pathogenesis of Diabetes Mellitus" (E. Cerasi and R. Luft, eds.), p. 57. Almqvist and Wiksell, Stockholm, (1970).
51. Steiner, D. F., Kemmler, W., Clark, J. L., Oyer, P. E., and Rubenstein, A. H., in "Handbook of Physiology - Endocrinology I, Chapter 10" (D. F. Steiner and N. Freinkel, eds.), p. 175. Williams, and Wilkins Co., Baltimore, (1972).
52. Steiner, D. F., Quinn, P. S., Patzelt, C., Chan, S. J., Marsh, J., and Tager, H. S., in "Cell Biology: A Comprehensive Treatise, Vol. IV" (Goldstein and Prescott, eds.), p. 175. Academic Press, New York, (1980).
53. Tager, H. S., Emdin, S. O., Clark, J. L., and Steiner, D. F., J. Biol. Chem. 248, 3476 (1973).
54. Takio, K., Towatari, T., Katanuma, N., Teller, D. C. and Titani, K., Proc. Natl. Acad. Sci. USA 80, 3666 (1983).
55. Virji, M. A. G., Vassalli, J-D., Estensen, R. D., and Reich, E., Proc. Natl. Acad. Sci. USA 77, 875 (1980).
56. Yoo, O. J., Powell, C. T. and Agarwal, K. L., Proc. Natl. Acad. Sci. USA 79, 1049 (1982).

PART I
Opioid Peptides

HUMAN β-LIPOTROPIN[1,2]

Choh Hao Li

Hormone Research Laboratory
University of California
San Francisco, CA

Human β-lipotropin (β_h-LPH) consists of 89 amino acids with glutamic acid as the NH_2- and COOH-terminal residue and its primary structure has been proposed. In order to securely establish the primary structure of β_h-LPH, the hormone has been synthesized by the new segment-coupling method in aqueous solution. The synthetic product was shown to be identical to native β_h-LPH by paper electrophoresis, isoelectric focusing, high performance liquid chromatography, lipolytic activity in isolated rabbit fat cells and radio-immunoassay. Since β-lipotropin is the prohormone for β-endorphin, some recent studies on β-endorphin are discussed.

β-Lipotropin was identified as a new hormone in 1964 from sheep pituitary glands (Li, 1964) and its amino acid sequence was reported in the following year (Li et al., 1965). It consists of 91 amino acids with glutamic acid and glutamine as NH_2 and COOH-terminal residues, respectively (see Figure 1). Total synthesis of sheep β-LPH was accomplished in 1978 (Yamashiro and Li, 1978). The synthetic β_s-LPH has been found to be indistinguishable from the natural hormone

[1]*Presented at the conference on "Biochemical and Clinical Aspects of Neuropeptides: Synthesis, Processing and Gene Structure"; Hamburg, May 29-June 2, 1983.*

[2]*This work was supported in part by National Institutes of Health Grant GM-2907, National Institute of Mental Health Grant MH-30245 and the Hormone Research Foundation.*

ISBN 0-12-417320-9

```
                                5                   10                  15                  20
  Human:  H-Glu-Leu-Thr-Gly-Gln-Arg-Leu-Arg-Glu-Gly-Asp-Gly-Pro-Asp-Gly-Pro-Ala-Asp-Asp-Gly-
  Ovine:  H-Glu-Leu-Thr-Gly-Glu-Arg-Leu-Glu-Gln-Ala-Arg-Gly-Pro-Glu-Ala-Gln-Ala-Glu-Ser-Ala-
 Bovine:  H-Glu-Leu-Thr-Gly-Glu-Arg-Leu-Glu-Gln-Ala-Arg-Gly-Pro-Glu-Ala-Gln-Ala-Glu-Ser-Ala-
Porcine:  H-Glu-Leu-Ala-Gly-Ala-Pro-Pro-Glu-Pro-Ala-Arg-Asp-Pro-Glu-Ala-Pro-Ala-Glu-Gly-Ala-
Ostrich:  H-Ala-Leu-Pro-Pro-Ala-Ala-Met-Leu-Pro-Ala-Ala-Ala-Glu-Glu-Glu-Glu-Gly-Glu-Glu-Glu-

                  25                  30                                         35
            Ala-Gly-Ala-Gln-Ala-Asp-Leu-Glu-His-Ser-Leu-Leu-Val(                )Ala-Ala-
            Ala-Ala-Arg-Ala(    )Glu-Leu-Glu-Tyr-Gly-Leu-Val-Ala-Glu(       )Ala-Glu-Ala-Ala-
            Ala-Ala-Arg-Ala(    )Glu-Leu-Glu-Tyr-Gly-Leu-Val-Ala-Glu-Ala-Glu-Ala-Glu-Ala-Ala-
            Ala-Ala-Arg-Ala(    )Glu-Leu-Glu-His-Gly-Leu-Val-Ala-Glu(       )Ala-Gln-Ala-Ala-
            Glu-Glu-Gly-Glu(                                                            )Ala-

                  40                  45                  50                  55
            Glu-Lys-Lys-Asp-Glu-Gly-Pro-Tyr-Arg-Met-Glu-His-Phe-Arg-Trp-Gly-Ser-Pro-Pro-Lys-
            Glu-Lys-Lys-Asp-Ser-Gly-Pro-Tyr-Lys-Met-Glu-His-Phe-Arg-Trp-Gly-Ser-Pro-Pro-Lys-
            Glu-Lys-Lys-Asp-Ser-Gly-Pro-Tyr-Lys-Met-Glu-His-Phe-Arg-Trp-Gly-Ser-Pro-Pro-Lys-
            Glu-Lys-Lys-Asp-Glu-Gly-Pro-Tyr-Lys-Met-Glu-His-Phe-Arg-Trp-Gly-Ser-Pro-Pro-Lys-
            Glu-Lys-Glu-Asp-Gly-Gly-Ser-Tyr-Arg-Met-Arg-His-Phe-Arg-Trp-Gln-Ala-Pro-Leu-Lys-

                  60                  65                  70                  75
            Asp-Lys-Arg-Tyr-Gly-Gly-Phe-Met-Thr-Ser-Glu-Lys-Ser-Gln-Thr-Pro-Leu-Val-Thr-Leu-
            Asp-Lys-Arg-Tyr-Gly-Gly-Phe-Met-Thr-Ser-Glu-Lys-Ser-Gln-Thr-Pro-Leu-Val-Thr-Leu-
            Asp-Lys-Arg-Tyr-Gly-Gly-Phe-Met-Thr-Ser-Glu-Lys-Ser-Gln-Thr-Pro-Leu-Val-Thr-Leu-
            Asp-Lys-Arg-Tyr-Gly-Gly-Phe-Met-Thr-Ser-Glu-Lys-Ser-Gln-Thr-Pro-Leu-Val-Thr-Leu-
            Asp-Lys-Arg-Tyr-Gly-Gly-Phe-Met-Ser-Ser-Glu-Arg-Gly-Arg-Ala-Pro-Leu-Val-Thr-Leu-

                  80                  85             89         Total Residues
            Phe-Lys-Asn-Ala-Ile-Ile-Lys-Asn-Ala-Tyr-Lys-Lys-Gly-Glu-OH        89
            Phe-Lys-Asn-Ala-Ile-Ile-Lys-Asn-Ala-His-Lys-Lys-Gly-Gln-OH        91
            Phe-Lys-Asn-Ala-Ile-Ile-Lys-Asn-Ala-His-Lys-Lys-Gly-Gln-OH        93
            Phe-Lys-Asn-Ala-Ile-Val-Lys-Asn-Ala-His-Lys-Lys-Gly-Gln-OH        91
            Phe-Lys-Asn-Ala-Ile-Val-Lys-Ser-Ala-Tyr-Lys-Lys-Gly-Gln-OH        79
```

FIGURE 1. Amino acid sequence of β-lipotropin from various species in its R_f value on partition chromatography, mobility in paper electrophoresis at two pH values, behavior on thin-layer chromatography, amino acid composition of both acid and enzymatic hydrolysates, NH_2-terminal residue, behaviour on peptide mapping, pI value in isoelectric focusing, circular dichroism spectra, optical rotation, lipolytic activity and immunoreactivity.

β-LPH has also been obtained in highly purified form from cow (Lohmar and Li, 1967), pig (Cseh et al., 1968; Gilardeau and Chrétien, 1970; Yudeau and Pankov, 1970), rat (Rubinstein et al., 1977), whale (Kawauchi et al., 1980), turkey (Chang et al., 1980) and ostrich (Naudé and Oelofsen, 1981) pituitary glands. The complete amino acid sequences of pig (Gráf et al., 1971; Gilardeau and Chrétien, 1972; Pankov and Yudeau, 1972), cow (Li et al., 1977; Nakanish et al., 1979), turkey (Chang et al., 1980) and ostrich (Naudé et al., 1981) β-LPH have been proposed. The following describes the isolation, primary structure and synthesis of human β-LPH.

Isolation

β-LPH has been isolated from human pituitary glands by Cseh et al. (1968) and other investigators (Scott and Lowry, 1974; Chrétien et al., 1976; Li and Chung, 1976). The isolation procedure was modified from that employed for the sheep hormone (Li and Chung, 1981). Briefly, 385 glands were homogenized in 75 ml distilled water for 1 min in a Waring Blender. The slurry was then added to cold 600 ml acetone plus 15 ml conc. HCl and the mixture stirred for 1 h. The mixture was filtered using Whatman No. 1 paper and the residue re-extracted with 300 ml 80% acetone by stirring for 1/2 h. The residue was filtered off and discarded. The combined supernatants were added slowly to 4.5 l of cold acetone with stirring for 5 min. The precipitate was allowed to settle for 2 days. The acetone was siphoned off and discarded. The precipitate was filtered, washed with cold acetone, and dried overnight in vacuo over NaOH pellets. The resulting acid-acetone powder (4.1 g) was dissolved in 125 ml of water at pH 3 and the solution was brought to 0.06 saturation with respect to NaCl. The NaCl precipitate was removed by centrifugation and the solution was brought to saturation with NaCl. The saturated NaCl precipitate was dissolved in 100 ml of 0.1 M HOAc and desalted by gel filtration on Sephadex G-10 column (3.5 x 80 cm) in 0.1 M HOAC. After lyophilization, the yield of the desalted material (designated fraction D) was 444 mg.

Fraction D was next submitted to chromatography on a carboxymethyl cellulose column. Further purification was performed with exclusion chromatography on Sephadex G-50 in 0.1 M HOAc. The content in the main peak was lyophilized, yielded 12 mg and was shown to be homogeneous in paper electrophoresis at pH 6.7 and disc electrophoresis. NH_2-terminal analysis gave only glutamic acid as the end group. This material was subsequently characterized as human β-LPH.

Primary Structure

From the tryptic, chymotryptic, peptic and thermolysin peptides of β_h-LPH, it is possible to propose the complete amino acid sequence of the hormone (Li and Chung, 1981). It consists of 89 amino acids with the single Trp residue in position 50.

The sequence of COOH-terminal 52 residues has been reported by Cseh et al. (1972). In addition, these investigations indicated that their β_h-LPH preparation has the NH_2-terminal sequence: NH_2-Glx-Gly-Asx-. The sequence as proposed (Li and Chung, 1981) is in agreement with Cseh et al. (1972) on the COOH-terminal sequence but differed on the sequence at the NH_2-terminus. It is of interest to note that Chrétien et al. (1976) reported NH_2-Glu-Leu-Thr-Gly-Glu- and -Ala-Val-Leu-OH as NH_2-terminal and COOH-terminal sequences respectively in β_h-LPH. In the meantime, Chang et al. (1980) reported the nucleotide sequence of the human proopiomelanocortin gene. From the nucleotide sequence, the amino acid sequence of β_h-LPH may be deduced. When the deduced sequence is compared with that obtained by the direct sequence analysis (Li and Chung, 1981), differences in residues for 11 positions at the NH_2-terminal sequence are evident. This led Hsi et al. (1981), Spiess et al. (1982) and Seidah et al. (1982) to reinvestigate the NH_2-terminal sequence of β_h-LPH. Results indicate that 5 residues in positions 9, 14, 15, 16 and 21 in the earlier proposed structure (Li and Chung, 1981) are misplaced:

```
   9         14  15  16        21
---Gln-------Asn-Ala-Gly------Pro---   (Li & Chung, 1981)

---Glu-------Asp-Gly-Pro------Ala---   (Hsi et al., 1981;
                                        Spiess et al.,
                                        1982; Sediah et
                                        al., 1982)
```

In addition, Gln-9 and Asn-14 earlier proposed (Li and Chung, 1981) were found to be Glu and Asp, respectively (Hsi et al., 1981; Spiess et al., 1982; Seidah et al., 1982). The revision is in agreement with the deduced amino acid sequence from the most recent nucleotide sequence of the human pro-hormone by Takahashi et al. (1981). In order to establish the primary structure of β_h-LPH, it was decided to synthesize the hormone according to the primary structure as shown in Figure 1. When compared with the primary structure of ovine, bovine, porcine or ostrich hormone, it is evident that the carboxyl terminal segment is highly conserved in evolution. Differences occur in the first NH_2-terminal 34-39 residues. It is of interest to note that the total number of residues varies from 79 in ostrich to 93 in bovine β-LPH.

Synthesis

The synthesis was carried out by the new segment-coupling method in aqueous solution as recently described by Blake (1981) and Blake and Li (1981). This involves the use of thiol acids for coupling segments in aqueous media as shown in the following reaction:

$$R_1CO\text{-}S^- + 2Ag^+ + H_2N\text{-}R_2 \longrightarrow R_1CONHR_2 + Ag_2S + H^+$$

Where R_1 and R_2 are peptide segments which may be synthesized by the solid-phase method.

The strategy for the total synthesis of β_h-LPH (Blake and Li, 1983) was to synthesize peptides corresponding to β_h-LPH-(1-60) and β_h-LPH-(61-89) which could be coupled together to give the Gly^{60}-Gly^{61} peptide bond. In order to obtain the desired peptide [Lys(Cit)^{67,77,82,86,87}]-β_h-LPH-(61-89) (Ia) by conventional procedures of solid-phase peptide synthesis it was necessary to block the terminal α-amino function with a group that would be stable to HF and thus allow selective protection of the lysine amino groups. Previously (Blake and Li, 1981), the 9-(fluorenyl)methyloxy-carbonyl group was useful for this purpose, but in this case the decrease in aqueous solubility conferred by this group was unacceptable. It was decided to take advantage of the absence of arginine from β_h-LPH-(61-89). Accordingly, Ac-Arg-β_h-LPH-(61-89) (I) was synthesized and readily purified by chromatography on CM-cellulose and partition chromatography on Sephadex G-50. Reaction of peptide I with citraconic anhydride, followed by digestion with trypsin to remove the acetylarginyl residue, gave Ia.

Boc-Gly-S resin (Blake and Li, 1981) was used for the synthesis of [GlyS^{60}]-β_h-LPH-(1-60) (II) by the solid-phase method. The crude product obtained from HF cleavage was chromatographed on CM-cellulose, deformylated by brief treatment with aqueous alkali, and rechromatographed on CM-cellulose. Partition chromatography gave the higly purified product which was treated with citraconic anhydride to give Cit-[Lys(Cit)^{37,38,55,57},GlyS^{60}]-β_h-LPH-(1-60) (IIa).

Peptides Ia and IIa were coupled together by reaction with silver nitrate/N-hydroxysuccinimide in 50% dimethylformamide. Overnight reaction with 25% acetic acid gave the crude product that was purified by chromatography on CM-cellulose. The material in the peak tubes which had the same elution position as native β_h-LPH, was shown to be the 89-peptide by amino acid analysis. Further purification was necessary by gel chromatography on Sephadex G-50 and rechromatography on CM-cellulose. The final yield of β_h-LPH was 10%. The synthetic product was identical to the natural hormone in its electrophoretic mobility on paper, behavior on HPLC (Figure 2), and position in isoelectric focusing gels. Assay for lipolytic activity in isolated rabbit fat cells and radioimmuoreactivity showed the synthetic product to be equipotent with the native hormone (Table I). Thus, the

Table I. Biological Activity of Synthetic β_h-LPH

	Lipolytic activity			*Immunoreactivity*	
β_h-LPH	*Dose (ng/ml)*	*Glycerol liberated[a] (μmol/g cell ml/hr)*	*Relative potency*	*IC_{50} fmol/ tube*	*Relative potency*
Native	*100*	*1.80 ± 0.13*	*100*	*52.9*	*100*
	300	*2.51 ± 0.20*		*(32.6-88.2)[b]*	
Synthetic	*100*	*1.89 ± 0.20*	*117[c]*	*39.7*	*133*
	300	*2.63 ± 0.07*		*(14.1-149.4)*	

[a]Mean of triplicate determinations.

[b]95% Confidence limit

[c]95% Confidence limit, 63-242 and index precision (λ), 0.18

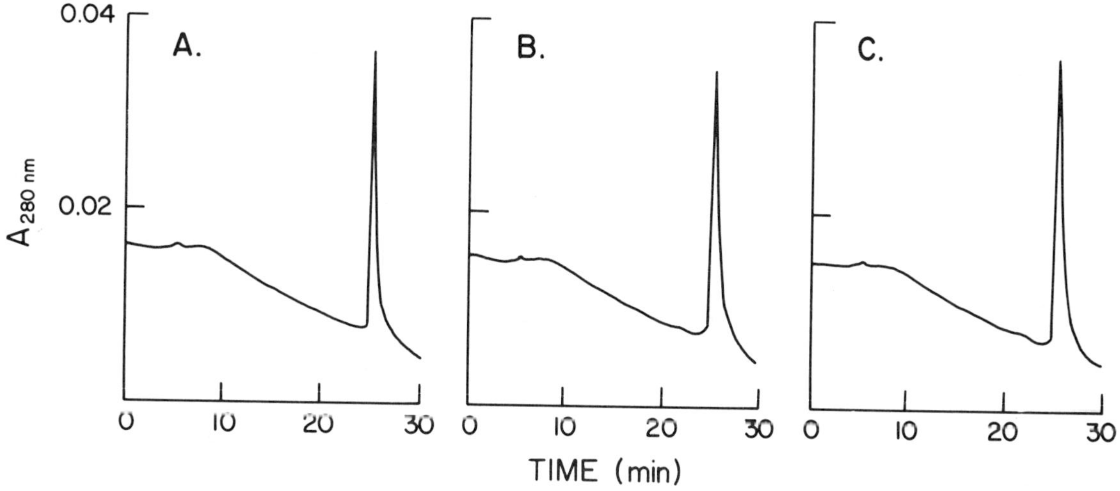

FIGURE 2. HPLC of 20-μg samples of native β_h-LPH (A) synthetic β-LPH (B) and a 1:1 mixture (C) on an Alltech Vydac 201-TP column (4.5 x 250 mm). Elution was by a linear gradient of 10-50% isopropanol in 0.1% trifluoroacetic acid over 30 min.

amino acid sequence as shown in Figure 1 is the correct primary structure for β_h-LPH. This illustrates the continuing importance of organic synthesis in structure elucidation.

β_h-Endorphin

β_h-Lipotropin is the prohormone for β-EP as it has been shown by Gráf et al. (1977) that a specific intracellular protease in the pituitary causes the cleavage of the Arg-Tyr bond in positions 60 and 61 in the porcine β-LPH to generate the opioid peptide. β-Endorphin has been isolated and sequenced from pituitary glands of various species. They all consist of 31 amino acids. From their amino acid sequence as shown in Figure 3, only 8 out of 31 residues in the ostrich structure differed from that of the human hormone. In the case of camel β-EP, it differs from the human hormone in only two positions:

```
          25        31
Camel:  -His--------Gln-OH
Human:  -Tyr--------Glu-OH
```

These slight differences in the primary structure are clearly evident in their binding behavior in two neuroblastoma cell lines (Hammonds et al., 1981; Hammonds and Li, 1981) using ^3H-β_h-EP (Houghten et al., 1980) as the primary ligand.

Specific binding of β_h-EP to NG108-15 cells is time dependent and saturable with $K_d = 0.3$ and $k_a = 1.8 \times 10^8$ $M^{-1}min^{-1}$. Under the conditions optimal for β_h-EP binding, Leu-enkephalin (Leu-EK) has one-fourth to one-third as many binding sites as β_h-EP and its affinity is 7-10% that of β_h-EP. [Leu^5]-β_h-EP is approximately equipotent with β_h-EP and capable of displacing all specific binding. β_c-EP displaces only 90% of the specific binding, yet is more potent than β_h-EP in displacing the remainder (see Figure 4). However, in the N18TG2 cells, the affinity and number of binding sites for β_h-EP are 1.1 nM and 280,000 per cell, respectively. β_h-EP binding is inhibited by Leu-EK or morphine; β_h-EP-(6-31) is a potent inhibition of binding and β_c-EP is much less potent (see Figure 5). The data suggest the importance of the non-EK segment of the β_h-EP molecule for interaction with the binding site in N18TG2 cells

Species	1 – 5 – 10 – 15	20 – 25 – 30
Human:	H-Tyr-Gly-Gly-Phe-Met-Thr-Ser-Glu-Lys-Ser-Gln-Thr-Pro-Leu-Val-Thr-Leu-	Phe-Lys-Asn-Ala-Ile-Ile-Lys-Asn-Ala-Tyr-Lys-Lys-Gly-Glu-OH
Camel, bovine, ovine:	H-Tyr-Gly-Gly-Phe-Met-Thr-Ser-Glu-Lys-Ser-Gln-Thr-Pro-Leu-Val-Thr-Leu	Phe-Lys-Asn-Ala-Ile-Ile-Lys-Asn-Ala-His-Lys-Lys-Gly-Gln-OH
Porcine:	H-Tyr-Gly-Gly-Phe-Met-Thr-Ser-Glu-Lys-Ser-Gln-Thr-Pro-Leu-Val-Thr-Leu-	Phe-Lys-Asn-Ala-Ile-Val-Lys-Asn-Ala-His-Lys-Lys-Gly-Gln-OH
Equine:	H-Tyr-Gly-Gly-Phe-Met-Ser-Ser-Glu-Lys-Ser-Gln-Thr-Pro-Leu-Val-Thr-Leu-	Phe-Lys-Asn-Ala-Ile-Ile-Lys-Asn-Ala-His-Lys-Lys-Gly-Gln-OH
Rat:	H-Tyr-Gly-Gly-Phe-Met-Thr-Ser-Glu-Lys-Ser-Gln-Thr-Pro-Leu-Val-Thr-Leu-	Phe-Lys-Asn-Ala-Ile-Ile-Lys-Asn-Val-His-Lys-Lys-Gly-Gln-OH
Turkey:	H-Tyr-Gly-Gly-Phe-Met-Thr-Ser-Glu-His-Ser-Gln-Met-Pro-Leu-Leu-Thr-Leu-	Phe-Lys-Asn-Ala-Ile-Val-Lys-Ser-Ala-Tyr-Lys-Lys-Gly-Gln-OH
Cstrich:	H-Tyr-Gly-Gly-Phe-Met-Ser-Ser-Glu-Arg-Gly-Arg-Ala-Pro-Leu-Val-Thr-Leu-	Phe-Lys-Asn-Ala-Ile-Val-Lys-Ser-Ala-Tyr-Lys-Lys-Gly-Gln-OH

FIGURE 3. Amino acid sequence of β-endorphin from various species

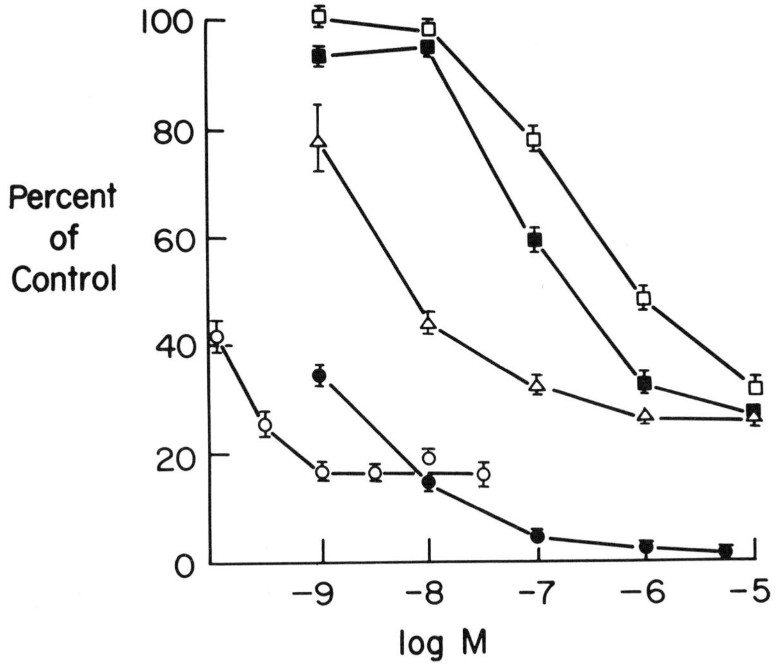

FIGURE 4. Competitive inhibition of ^{3}H-β_h-EP binding to NG108-15 cells by morphine (□), naloxone (■), [Leu]EK (Δ), β_h-EP (●), and β_c-EP (O). The behavior of [Met]EK was identical to that of [Leu]EK.

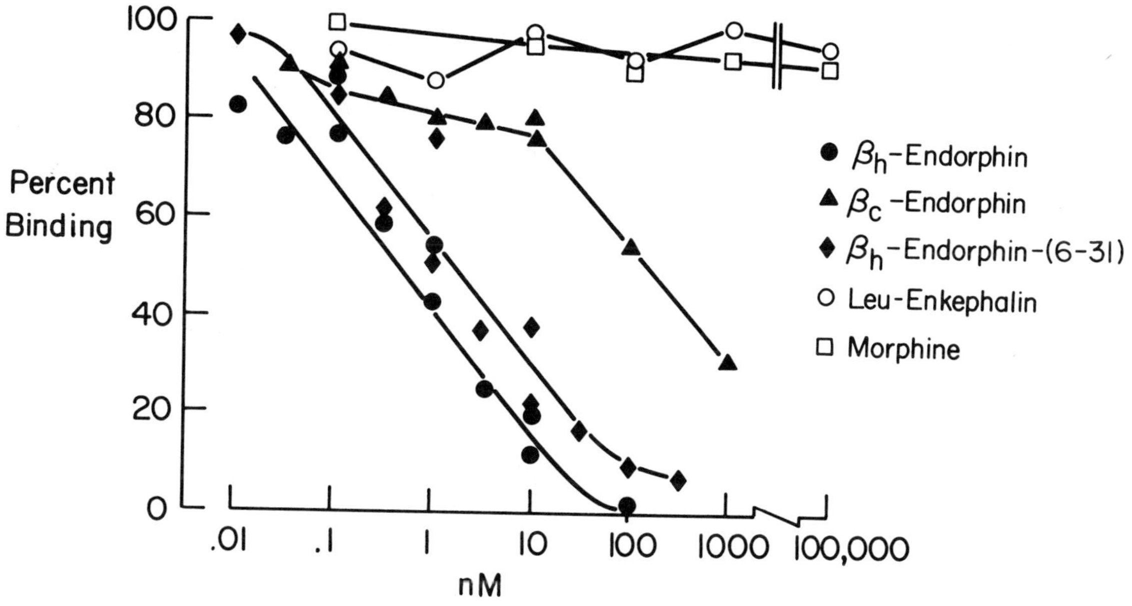

FIGURE 5. Inhibition of ^3H-β_h-EP binding to N18TG2 cells. N18TG2 cells and ^3H-β_h-EP (0.8 nM) were incubated with increasing amounts of the indicated compounds. Percent binding was calculated as 100 x $(B_s-B_n)/(B_o-B_n)$, in which B_o is the amount bound in the absence of competing compounds, B_s is the amount bound in the presence of a concentration x of competing compound, and B_n is the nonspecific binding, in this case the binding in the presence of 1 μM β_h-EP. The apparent IC_{50} values (concentration for 50% inhibition) are β_h-EP, 0.6 nM; β_h-EP-(6-31), 2.5 nM; β_c-EP, 200 nM; morphine and EK, >100 μM.

Table II. Analgesic Potency of Human, Camel, Equine, Turkey and Ostrich β-Endorphins

Peptides	*AD_{50}, nmol/mouse*[a]	*Relative potency*[b]
β_h-EP	0.033 (0.021-0.052)	100
β_c-EP	0.020 (0.014-0.028)	165
β_e-EP	0.021 (0.013-0.037)	153
β_{tu}-EP	0.072 (0.048-0.107)	45
β_{os}-EP	0.030 (0.020-0.045)	110

[a] *AD_{50} is the mean analgesic dose by intracerebroventricular injection. The numbers in parentheses refer to the 95% confidence limits.*

[b] *Relative potencies are calculated on a molar basis using the AD_{50} of β_h-EP as 100%.*

Table III. Radioreceptor Binding Activities of Human, Camel, Equine, Turkey and Ostrich β-Endorphins

Peptides	*IC_{50}, nM*[a]	*Relative potency*[b]
β_h-EP	0.73 ± 0.08	1.00 ± 0.16
β_c-EP	0.27 ± 0.02	2.70 ± 0.36
β_e-EP	0.22 ± 0.01	3.30 ± 0.40
β_{tu}-EP	0.76 ± 0.08	0.96 ± 0.15
β_{os}-ep	0.12 ± 0.01	6.10 ± 0.80

[a] *Fifty percent inhibitory concentration in the opiate receptor binding assay.*

[b] *Relative potencies are calculated on a molar basis using the IC_{50} of β-EP as 100%.*

Equine (Li et al., 1981), turkey (Yamashiro et al., 1980) and ostrich (Yamashiro et al., 1982) have been synthesized in addition to that of camel (Li et al., 1976) and human (Li et al., 1977) hormone. Analgesic potencies of these synthetic β-EPs have been determined in the mouse tail-flick test and binding affinities estimated by radioreceptor assay using rat brain membranes (Hammonds et al., 1982). Results are summarized in Tables II and III. The analgesic potency and binding affinity of β_c-EP are 1.7 and 2.7 times higher than the human hormone, respectively. His-27 and/or Gln-31 may contribute to this increased activity. The dissociation of radioreceptor binding affinity from analgesic potency in these naturally occurring β-EP homologs suggests that either the conditions under which the binding assay is performed make the true binding potency in the brain, or once bound to the appropriate receptor, these homologs do not possess equal ability to produce biological effects.

ACKNOWLEDGMENTS

I thank my coworkers whose names appear in the cited references for very pleasant collaborations.

REFERENCES

Blake, J., *Int. J. Peptide Protein Res. 17*, 273 (1981).
Blake, J., and Li, C. H., *Proc. Natl. Acad. Sci. USA 78*, 4055 (1981).
Blake, J., and Li, C. H., *Proc. Natl. Acad. Sci. USA 80*, 1556 (1983).
Chang, A.C.Y., Cochet, M., and Cohen, S. N., *Proc. Natl. Acad. Sci. USA 77*, 4890 (1980).
Chang, W-C., Chung, D., and Li, C. H., *Int. J. Peptide Protein Res. 15*, 261 (1980).
Chrétien, M., Gilardeau, C., Seidah, N., and Lis, M., *Canad. J. Biochem. 54*, 778 (1976).
Cseh, G., Gráf, L., and Both, F., *FEBS Lett. 2*, 42 (1968).
Cseh, G., Barat, E., Patthy, A., and Gráf, L., *FEBS Lett. 21*, 344 (1972).
Gilardeau, C., and Chrétien, M., *Canad. J. Biochem. 48*, 1017 (1970).
Gilardeau, C., and Chrétien, M., *in* "Chemistry and Biology of Peptides" (J. Meienhofer, ed.), p. 609. Ann Arbor Science Publ., Ann Arbor, MI (1972).
Gráf, L., Barat, E., Cseh, G., and Sajco, M., *Biochim. Biophys. Acta 229*, 276 (1971).
Gráf, L., Kenessey, A., Berzétol, I., and Rónai, A. Z., *Biochem. Biophys. Acta 78*, 1114 (1977).

Hammonds, R. G., Jr., and Li, C. H., *Proc. Natl. Acad. Sci. USA 78*, 6764 (1981).
Hammonds, R. G., Jr., Ferrara, P., and Li, C. H., *Proc. Natl. Acad. Sci. USA 78*, 2218 (1981).
Hammonds, R. G., Jr., Nicolas, P., and Li, C. H., *Int. J. Peptide Protein Res. 19*, 556 (1982).
Houghten, R. A., Chang, W-C., and Li, C. H., *Int. J. Peptide Protein Res. 16*, 311 (1980).
Hsi, K. L., Seidah, N. G., Lu, C. H., and Chrétien, M., *Biochem. Biophys. Res. Commun. 103*, 1329 (1981).
Kawauchi, H., Chung, D., and Li, C. H., *Int. J. Peptide Protein Res. 15*, 171 (1980).
Li, C. H., *Nature 201*, 924 (1964).
Li, C. H., and Chung, D., *Nature 260*, 623 (1976).
Li, C. H., and Chung, D., *Int. J. Peptide Protein Res. 17*, 131 (1981).
Li, C. H., Barnafi, L., Chrétien, M., and Chung, D., *Nature 208*, 1093 (1965).
Li, C. H., Lemaire, S., Yamashiro, D., and Doneen, B. A., *Biochem. Biophys. Res. Commun. 71*, 19 (1976).
Li, C. H., Tan, L., and Chung, D., *Biochem. Biophys. Res. Commun. 77*, 1088 (1977).
Li, C. H., Yamashiro, D., Tseng, L-F., and Loh, H. H., *J. Med. Chem. 20*, 325 (1977).
Li, C. H., Ng, T. B., Yamashiro, D., Chung, D., Hammonds, R. G., Jr., and Tseng, L-F., *Int. J. Peptide Protein Res. 18*, 242 (1981).
Lohmar, P., and Li, C. H., *Biochim. Biophys. Acta 147*, 381 (1967).
Nakanishi, S., Inoue, A., Kita, T., Nakamura, M., Chang, A.C.Y., Cohen, S. N., and Numa, S., *Nature 278*, 423 (1979).
Naudé, R. J., and Oelofsen, W., *Int. J. Peptide Protein Res. 18*, 135 (1981).
Naudé, R. J., Chung, D., Li, C. H., and Oelofsen, W., *Int. J. Peptide Protein Res. 18*, 138 (1981).
Pankov, Y. A., and Yudeau, M. A., *Biokimia 37*, 991 (1972).
Rubinstein, M., Stein, S., Gerber, L., and Udenfriend, S., *Proc. Natl. Acad. Sci. USA 74*, 3052 (1977).
Scott, A. P., and Lowry, P. J., *Biochem. J. 139*, 593 (1974).
Seidah, N. G., Hsi, K. L., Chrétien, M., Barat, E., Patthy, A., and Gráf, L., *FEBS Lett. 147*, 267 (1982).
Speiss, J., Mount, C. D., Nicholson, W. E., and Orth, D. N., *Proc. Natl. Acad. Sci. USA 79*, 5071 (1982).
Takahashi, H., Teranishi, Y., Nankanishi, S., and Numa, S., *FEBS Lett. 135*, 97 (1981).
Yamashiro, D., and Li, C. H., *J. Amer. Chem. Soc. 100*, 5174 (1978).

Yamashiro, D., Ferrara, P., and Li, C. H., *Int. J. Peptide Protein Res. 16*, 75 (1980).
Yamashiro, D., Hammonds, R. G., Jr., and Li, C. H., *Int. J. Peptide Protein Res. 19*, 251 (1982).
Yudeau, M. A., and Pankov, Y. A., *Endocrinology (USSR) 16*, 49 (1970).

BIOCHEMICAL STUDIES OF PROENKEPHALIN AND THE PRODUCTS DERIVED FROM ITS PROCESSING

Sidney Udenfriend

Roche Institute of Molecular Biology
Roche Research Center
Nutley, New Jersey

Experiments that led to the discovery of proenkephalin and the products of its processing are presented. The physiologic significance of the many EC-peptides derived from proenkephalin is discussed.

The search for proenkephalin began when we realized that β-endorphin, the enkephalin-containing peptide (EC-peptide) derived from pro-opiomelanocortin, did not give rise to [Met]- or [Leu]enkephalin (1). Evidence for a multienkephalin-containing protein was first obtained on extracts of guinea pig brains (2). Subsequently when beef adrenal medulla was reported to be a rich source of immunoreactive enkephalin (3) our attention turned to this tissue. EC peptides, ranging in size from ∿ 1000 daltons to ∿ 20,000 daltons were found in acid extracts of medulla and a still larger EC peptide(s) was observed in neutral extracts (4). Of greater interest was the fact that even the larger EC peptides contained several copies of enkephalin. In fact the largest peptide appeared to contain from 6 to 7 [Met]enkephalin sequences as well as one of [Leu]-enkephalin. This report in 1980 (4) represented the first evidence of the unusual gene product proenkephalin.

With the micromethodology for peptides and proteins that we had developed in our laboratories (5) we began the systematic purification and characterization of all the EC peptides that we found in bovine adrenal gland. The chromaffin granules, in which the EC peptides are stored were first isolated by sedimentation procedures, thereby achieving a many fold purification even prior to extraction. Extraction of the granules was then carried out with a dilute acid solution containing protease inhibitors to minimize proteolysis. Several steps of

BIOCHEMICAL AND CLINICAL ASPECTS OF
NEUROPEPTIDES: SYNTHESIS, PROCESSING,
AND GENE STRUCTURE

ISBN 0-12-417320-9

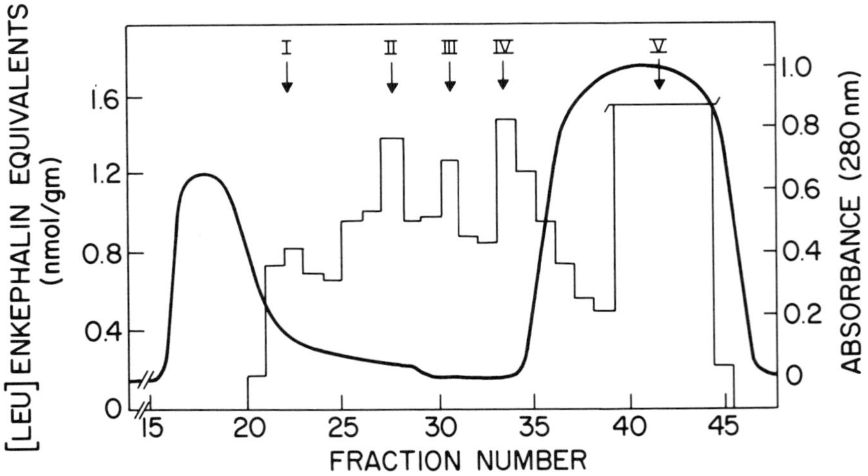

FIGURE 1. G-75 chromatography of chromaffin granule extracts. The bars represent enkephalin immunoreactivity following digestion with trypsin and carboxypeptidase B. Each peak of activity, labelled by a roman numeral, comprised a mixture of peptides that was resolved into the components shown in Table I.

chromatography were carried out to purify the approximately sixteen EC peptides to homogeneity. At each chromatographic step EC peptides in the eluates were treated with trypsin and carboxypeptidase B to release free enkephalins from their bound form and the latter were assayed by either radioimmunoassay or receptor binding assay. The G-75 chromatogram, shown in Figure 1, demonstrates the complex mixture of EC peptides that we found in adrenal extracts. Fractions eluted from such a chromatogram were arbitrarily divided according to size and each subjected to several steps of HPLC using a variety of supports and gradients. The overall purification of one of the larger EC-peptides, 8.6 kDa, is shown in Figure 2. Similar methods were used to purify the sixteen EC peptides. Criteria of homogeneity were attainment of constant specific activity, the appearance of a single sharp peak on two successive HPLC runs and, for the larger polypeptides, a single band on SDS gel electrophoresis.

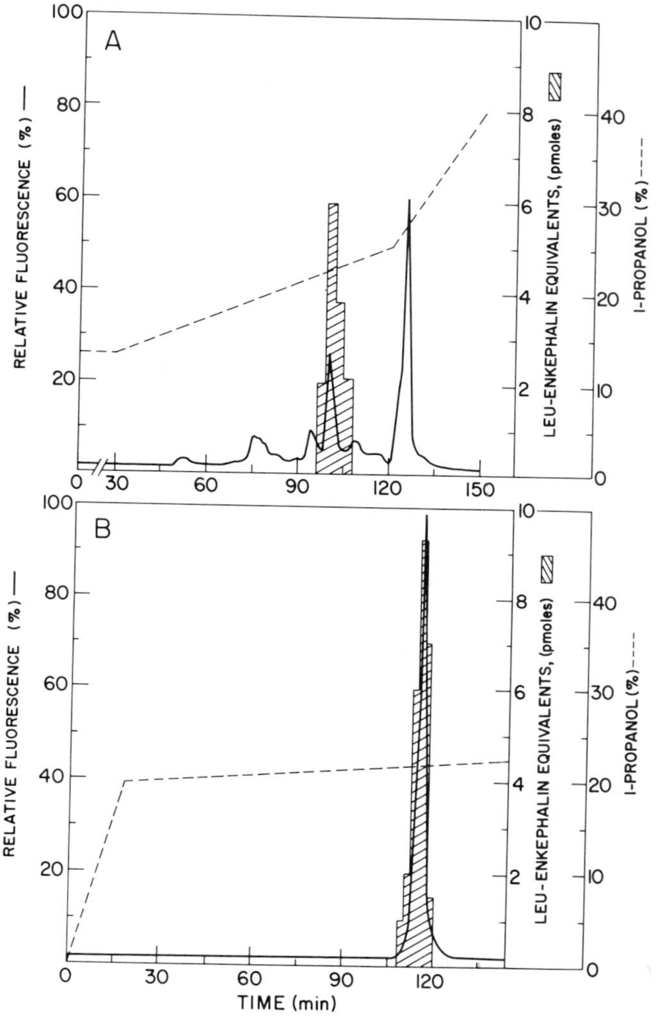

FIGURE 2. Purification of the 8.6 kDa ECP.

Early on it was decided to sequence the smaller peptides completely but to limit sequencing of the larger ones to as long stretches of the amino and carboxyl termini as possible. At that time we used the modified spinning cup method for amino terminal sequencing (6). One to two nanomoles were required and stretches up to 20 to 30 sequences could be obtained. Only 200 to 300 pmol of peptide were required for sequencing from the carboxyl terminus using carboxypeptidase Y to generate amino acids sequentially and prelabelling with *o*-phthalaldehyde to assay the liberated amino acids (7).

Table I lists all the EC peptides that were isolated from bovine adrenal medulla and the sequence data that was available prior to cloning. From this one can already deduce a large portion of the proenkephalin structure and the pathway of its processing. It is obvious that the three largest peptides, 18.2, 12.6 and 8.6 kDa, were different sized fragments of the same precursor. The fact that all three were identical from their amino termini already made it likely that the amino terminus of proenkephalin began with the sequence Glu-Cys-Ser-Gln-Asp-. We had predicted earlier that the heptapeptide [Met]-enkephalin -Arg^6-Phe^7 represented the carboxyl terminus of proenkephalin (8). If this were so then Peptide B, with the heptapeptide at its carboxyl terminus, should represent a longer stretch of the carboxyl terminus of proenkephalin. Discovery of the internal sequence -Tyr-Gly-Gly-Phe-Met-Arg-Gly-Leu- in two of the larger EC peptides suggested to us that the octapeptide Tyr-Gly-Gly-Phe-Met-Arg-Gly-Leu was processed out intact and could have unusual activity (9). This was confirmed even before the entire structure of proenkephalin was deduced (10).

The EC peptides in Table I could be compared to the pieces of a jigsaw puzzle with some of the smaller connecting pieces still missing. To deduce their proper alignment and how they were joined in proenkephalin it was necessary to turn to recombinant DNA technology. Actually this had already been started while sequencing was going on. As soon as the structures of Peptide I (11) and BAM 22 (12) were elucidated it became obvious that it was possible to prepare a relatively large synthetic deoxynucleotide probe with essentially no degeneracy (Figure 3). The synthetic decahexamer was first used to demonstrate and characterize proenkephalin mRNA in beef adrenal extracts (13) and then to produce, clone, and sequence bovine medulla proenkepahlin cDNA (14). Numa and his colleagues used this sequence information to construct the same probe that enabled them to clone and sequence bovine medulla preproenkephalin cDNA. Identical sequence data were obtained by us and by the group in Kyoto (15) (Figure 4). As we had predicted (8) the heptapeptide [Met]enkephalin -Arg^6-Phe^7 represented the carboxyl terminus of proenkephalin. Information from the adjacent nucleotide sequences led Numa and his colleagues to select as the amino terminus of proenkephalin the sequence Glu-Cys-Ser-Gln- which represents the common amino termini of the three EC-peptides 18.2, 12.6 and 8.6 kDa. In our studies a combination of protein and cDNA sequencing led to the unequivocal selection of the correct proenkephalin amino terminus.

TABLE I Structural Data Of Isolated Adrenal Peptides Obtained By Protein Sequencing Methods. Parentheses Indicate Regions Which Were Not Sequenced.

PEAK V

Tyr-Gly-Gly-Phe-Met
Tyr-Gly-Gly-Phe-Met-Lys
Tyr-Gly-Gly-Phe-Met-Arg
Tyr-Gly-Gly-Phe-Met-Arg-Arg
Tyr-Gly-Gly-Phe-Met-Arg-Phe
Tyr-Gly-Gly-Phe-Met-Arg-Gly-Leu
Tyr-Gly-Gly-Phe-Leu
Tyr-Gly-Gly-Phe-Leu-Lys

PEAK IV

Peptide F

1 5 10 15 20 25 30
Tyr-Gly-Gly-Phe-Met-Lys-Lys-Met-Asp-Glu-Leu-Tyr-Pro-Leu-Glu-Val-Glu-Glu-Glu-Ala-Asn-Gly-Gly-Glu-Val-Leu-Gly-Lys-Arg-Tyr-Gly-Gly-Phe-Met

Peptide I

1 5 10 15 20 25 30
Ser-Pro-Thr-Leu-Glu-Asp-Glu-His-Lys-Glu-Leu-Gln-Lys-Arg-Tyr-Gly-Gly-Phe-Met-Arg-Arg-Val-Gly-Arg-Pro-Glu-Trp-Trp-Met-Asp-
35
Tyr-Gln-Lys-Arg-Tyr-Gly-Gly-Phe-Leu

Peptide B

1 5 10 25 30
Phe-Ala-Glu-Pro-Leu-Pro-Ser-Glu-Glu-Glu-Gly(Ser,Glx$_4$,Pro,Val,Met,Tyr,Lys)Lys-Arg-Tyr-Gly-Gly-Phe-Met-Arg-Phe

Peptide E

1 5 10 15 20 25
Tyr-Gly-Gly-Phe-Met-Arg-Arg-Val-Gly-Arg-Pro-Glu-Trp-Trp-Met-Asp-Tyr-Gln-Lys-Arg-Tyr-Gly-Gly-Phe-Leu

PEAK III

8.6-kDal EC-Peptide

1 5 10 15 20 25 30
Glu-Cys-Ser-Gln-Asp-Cys-Ala-Thr-Cys-Ser-Tyr-Arg-Leu-Ala-Arg-Pro-Thr-Asp-Leu-Asn-Pro-Leu-Ala-Cys-Thr-Leu-Glu-Cys-Glu-Gly-
35
Lys-Leu-Pro-Ser-Leu-Lys(Thr$_2$,Glx,Cys,Trp,Lys)(Thr,Glx$_2$,Leu$_3$,Lys)(Asx,Thr,Ser$_2$,Glx,Pro$_2$,Ala$_2$,Leu$_3$,Lys)(Ser,Glx$_3$,Ala,Leu$_2$,His)
75
Lys-Lys-Tyr-Gly-Gly-Phe-Met

PEAK II

5.3-kDal EC-Peptide

1 5 10 15 20 25 30
Asp-Ala-Glu-Glu-Asp-Asp-Gly-Leu-Gly-Asn-Ser-Ser-Asn-Leu-Leu-Lys-Glu-Leu-Leu-Gly-Ala-Gly-Asp-Gln-Arg-Glu-Gly-Ser-Leu-His-
35 40 45 50
Gln-Glu-Gly(Ser)Asp-Ala-Glu-Asp-Val(Ser)Lys(Arg)Tyr-Gly-Gly-Phe-Met-Arg-Gly-Leu

12.6-kDal EC-Peptide

1 5 10 15 20 25 30
Glu-Cys-Ser-Gln-Asp-Cys-Ala-Thr-Cys-Ser-Tyr-Arg-Leu-Ala-Arg-Pro-Thr-Asp-Leu-Asn-Pro-Leu-Ala-Cys-Thr-Leu-Glu-Cys-Glu-Gly-
35
Lys-Leu-Pro-Ser-Leu-Lys(Thr$_2$,Glx,Cys,Trp,Lys)(Thr,Glx$_2$,Leu$_3$,Lys)(Asx,Thr,Ser$_2$,Glx,Pro$_2$,Ala$_2$,Leu$_3$,Lys)(Ser,Glx$_3$,Ala,Leu$_2$,His)Ly
75 80 85 90 95 100
Lys-Tyr-Gly-Gly-Phe-Met-Lys-Arg-Tyr-Gly-Gly-Phe-Met-Lys-Lys-Met-Asp-Glu-Leu-Tyr-Pro-Leu-Glu-Val-Glu-Glu-Glu-Ala-Asn-Gly-Gly-
105 110
Glu-Val-Leu-Gly-Lys-Arg-Tyr-Gly-Gly-Phe-Met

PEAK I

18.2-kDal EC-Peptide

1 5 10 15 20 25 30
Glu-Cys-Ser-Gln-Asp-Cys-Ala-Thr-Cys-Ser-Tyr-Arg-Leu-Ala-Arg-Pro-Thr-Asp-Leu-Asn-Pro-Leu-Ala-Cys-Thr-Leu-Glu-Cys-Glu-Gly-
35
Lys-Leu-Pro-Ser-Leu-Lys(Thr$_2$,Glx,Cys,Trp,Lys)(Thr,Glx$_2$,Leu$_3$,Lys)(Asx,Thr,Ser$_2$,Glx,Pro$_2$,Ala$_2$,Leu$_3$,Lys)(Ser,Glx$_3$,Ala,Leu$_2$,His)
75 80 85 90 95 100
Lys-Lys-Tyr-Gly-Gly-Phe-Met-Lys-Arg-Tyr-Gly-Gly-Phe-Met-Lys-Lys-Met-Asp-Glu-Leu-Tyr-Pro-Leu-Glu-Val-Glu-Glu-Glu-Ala-Asn-Gly-
105 110 115 120 125 130
Gly-Glu-Val-Leu-Gly-Lys-Arg-Tyr-Gly-Gly-Phe-Met(Lys,Lys)Asp-Ala-Glu-Glu-Asp-Asp-Gly-Leu-Gly-Asn-Ser-Ser-Asn-Leu-Leu-Lys-Glu-
135 140 145 150 155 160
Leu-Leu-Gly-Ala-Gly-Asp-Gln-Arg-Glu-Gly-Ser-Leu-His-Gln-Glu-Gly(Ser)Asp-Ala-Glu-Asp-Val(Ser)Lys(Arg)Tyr-Gly-Gly-Phe-Met-Arg-
165
Gly-Leu

SEQUENCE WITHIN PEPTIDE E ---TRP-TRP-MET-ASP-TYR-GLN---

DNA PRIMER 3′ ACC-ACC-TAC-CTG-ATG-G 5′

FIGURE 3. Synthetic deoxynucleotide probe based on the unique sequence in Peptide I.

1 10 20 30
..//... Thr Val Arg Ala GLU CYS SER GLN ASP CYS ALA THR CYS SER TYR ARG LEU ALA ARG PRO THR ASP LEU ASN PRO LEU ALA CYS THR LEU GLU CYS GLU GLY
..//... ACC GUC AGG GCA GAA UGC AGC CAG GAC UGC GCC ACG UGC AGC UAC CGC CUG GCG CGC CCG ACU GAC CUC AAC CCG CUG GCU UGC ACU CUG GAA UGU GAG GGG

40 50 60
LYS LEU PRO SER LEU LYS Thr Trp Glu Thr Cys Lys Glu Leu Leu Gln Leu Thr Lys Leu Glu Leu Pro Pro Asp Ala Thr Ser Ala Leu
AAA CUA CCU UCU CUC AAG ACC UGG GAA ACC UGC AAG GAG CUU CUG CAG CUG ACC AAA CUA GAA CUU CCU CCA GAU GCC ACC AGU GCC CUC

70 80 90
Ser Lys Gln Glu Glu Ser His Leu Leu Ala LYS LYS TYR GLY GLY PHE MET LYS ARG TYR GLY GLY PHE MET LYS LYS MET ASP GLU LEU
AGC AAA CAG GAG GAG AGC CAC CUG CUU GCU AAG AAG UAC GGG GGC UUC AUG AAG CGG UAU GGG GGC UUC AUG AAG AAA AUG GAU GAG CUG

100 110 120
TYR PRO LEU GLU VAL GLU GLU GLU ALA ASN GLY GLY GLU VAL LEU GLY LYS ARG TYR GLY GLY PHE MET Lys Lys ASP ALA GLU GLU ASP
UAC CCC CUG GAA GUG GAA GAA GAG GCA AAU GGA GGU GAG GUC CUU GGC AAG AGA UAU GGG GGC UUC AUG AAG AAG GAU GCA GAG GAA GAU

130 140 150
ASP GLY LEU GLY ASN SER SER ASN LEU LEU LYS GLU LEU LEU GLY ALA GLY ASP GLN ARG GLU GLY SER LEU HIS GLN GLU GLY SER ASP
GAC GGC CUG GGC AAC UCC UCC AAC CUG CUC AAG GAG CUG CUG GGA GCC GGG GAC CAG CGA GAG GGG AGC CUC CAC CAG GAG GGC AGU GAU

160 170 180
ALA GLU ASP VAL SER LYS ARG TYR GLY GLY PHE MET ARG GLY LEU Lys Arg SER PRO HIS LEU GLU ASP GLU THR LYS GLU LEU GLN LYS
GCU GAA GAC GUG AGC AAG AGA UAC GGG GGC UUC AUG AGA GGC UUA AAG AGA AGC CCC CAC CUA GAA GAU GAA ACC AAA GAG CUG CAG AAG

190 200 210
ARG TYR GLY GLY PHE MET ARG ARG VAL GLY ARG PRO GLU TRP TRP MET ASP TYR GLN LYS ARG TYR GLY GLY PHE LEU Lys Arg PHE ALA
CGA UAC GGG GGU UUC AUG AGA AGA GUG GGU CGU CCA GAG UGG UGG AUG GAC UAC CAG AAA AGG UAC GGU GGC UUC CUC AAG CGC UUC GCC

220 230 240
GLU PRO LEU PRO SER GLU GLU GLU GLY Glu Ser Tyr Ser Lys Glu Val Pro Glu Met Glu LYS ARG TYR GLY GLY PHE MET ARG PHE
GAG CCC CUA CCC UCC GAG GAA GAA GGC GAA AGU UAC UCC AAG GAA GUU CCU GAA AUG GAG AAA AGA UAU GGA GGA UUU AUG AGA UUU UAA...

FIGURE 4. Nucleotide and amino acid sequences of bovine adrenal proenkephalin. Data obtained by protein sequencing are in bold type. Residues assigned solely on the basis of cDNA sequencing are in small type. Nucleotide sequences extending into the preproenkephalin region as determined by Noda et. al. *(15) are underlined.*

A diagramatic presentation of proenkephalin and the products of its processing that have been characterized is shown in Figure 5. The question remains as to what are the physiological products of processing? Clearly [Met]enkephalin-Arg^{6}-Phe^{7} and [Met]enkephalin-Arg^{6}-Gly^{7}-Leu^{8} are processed out intact, each sequence in proenkephalin being bracketed by paired basic amino acid residues. Both EC peptides have been shown to be more active than the free enkephalins in at least one assay system (8, 10). Because the remaining enkephalin sequences in proenkephalin are bracketed by double basic residues the Kyoto group

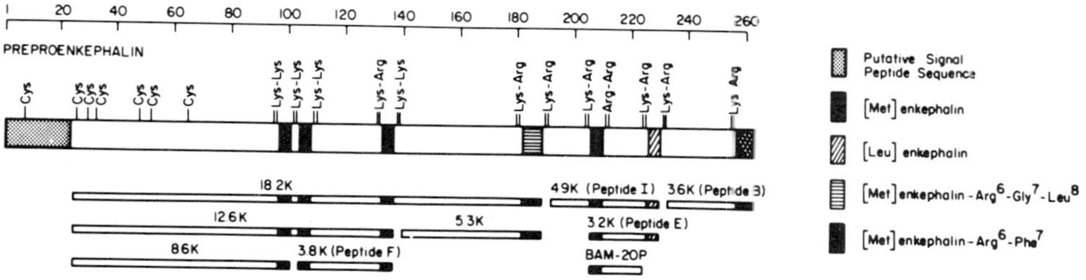

FIGURE 5. Schematic presentation of proenkephalin and the characterized products of its processing.

concludes that they are processed to four residues of [Met]enkephalins and one of [Leu]enkephalin (15). Theoretically this may be expected but this is not true, at least in the adrenal medulla. If one measures each of the EC peptides in extracts of adrenal chromaffin granules free enkephalins account for not more than 5 percent of the total (4). On stimulation the granules secrete an extract of the same composition (16). It would appear that the larger EC peptides are the products of processing, at least in this tissue.

It should be pointed out that Peptide E (Table II) is far more potent then the free enkephalins when tested on the guinea pig ileum (17). Like dynorphin (18) the enkephalin sequence of Peptide E is followed by -Arg-Arg-. It appears that this double basic residue after an enkephalin sequence is stable to the processing enzyme(s) which prefers -Lys-Lys- sequences. It should be noted that one of the heptapeptides isolated from adrenal extracts was [Met]enkephalin-Arg-Arg. There may be some significance to this since all enkephalin sequences, whether in proenkephalin or in pronorphin (prodynorphin), that are followed by -Arg-Arg- are processed out as larger EC peptides with -Arg-Arg- as an internal sequence. Furthermore all such compounds, Peptide E (17), dynorphin (18), α-neo-endorphin (19) and rimorphin (20) or dynorphin B(21) are far more active than the free enkephalins. Thus, in proenkephalin three of the enkephalin sequences are definitely intended to remain as part of larger and more active EC peptides. Even the four remaining enkephalin sequences of proenkephalin are still mainly present in the form of larger EC peptides when they are secreted from adrenal chromaffin granules. This is not peculiar for

TABLE II Opiate Activities of Enkephalin-Containing Peptides

Peptide	ED_{50} *(nmol)*[a] *Analgesia*	IC_{50} *(nM)*[b] *Guinea pig ileum assay*
[Met]enkephalin	*330.0*	*36*
[Leu]enkephalin	---	*504*
Peptide E	---	*0.9*
Bam-20P	---	*2.2*
[Met]enkephalin-Arg^6-Phe^7	*38.0*	*23*
[Met]enkephalin-Arg^6-Gly^7-Leu^8	---	*12*
β-Endorphin	*0.03*	*31*
Dynorphin	---	*0.3*
α-Neo-Endorphin	---	*5.6*
Rimorphin	---	*3.0*

[a] *Analgesia was determined in mice after intracerebral ventricular administration of the drug. The ED_{50} values represent nmol/mouse.*

[b] *While the IC_{50} values vary from laboratory to laboratory, the relative potencies of the peptides presented are in general agreement with observations reported by others.*

bovine adrenal medulla. As shown in Figure 6 guinea pig adrenal gland and intestine have a distribution of EC peptides similar to the bovine adrenal gland (22). Guinea pig and rat brain contain a larger proportion of free enkephalins but even in these tissues the EC peptides represent a large proportion of the total enkephalins present. EC peptides extended by sequences that are not readily cleaved by processing enzymes may represent the physiologically significant opioid agonists. It should be noted that the [Met]enkephalin sequence in β-endorphin is followed by a large extension beginning with -Ser-Thr- and is thus processed out as a large EC peptide that has high opiate activity even in vivo (23).

The discovery of the two enkephalins appeared to offer a simple mechanism for mediating opiate activity. However, the complexity of the problem has increased year by year. There are now known to be three genes that code

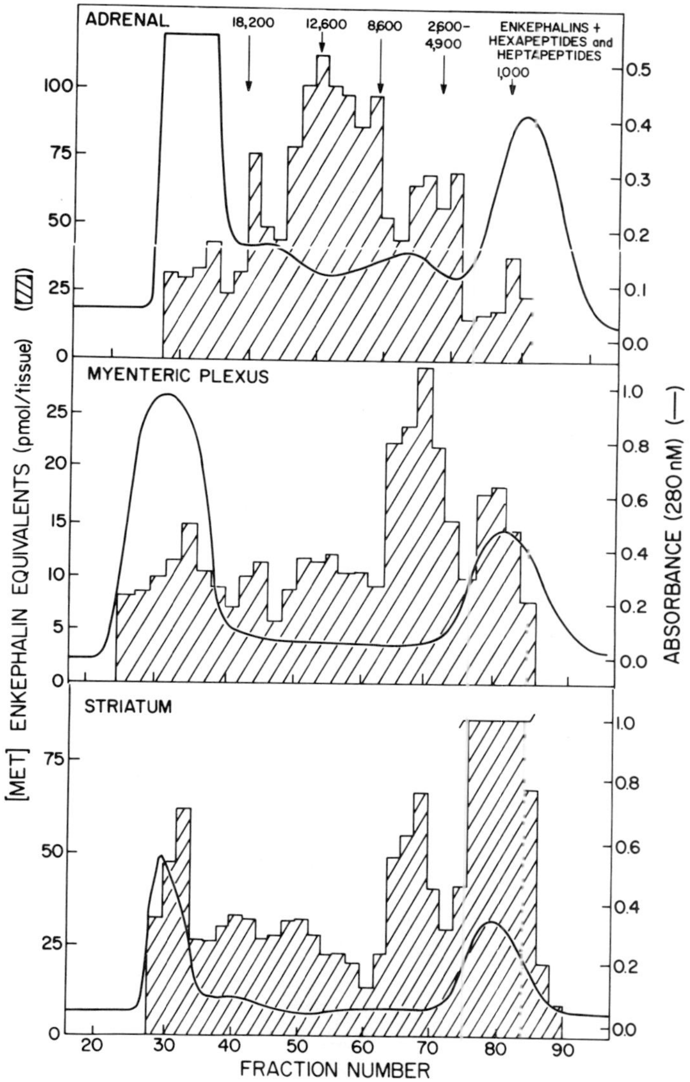

FIGURE 6. Chromatography of extracts of guinea pig tissues. The bars represent enkephalin immunoreactivity following digestion with trypsin and carboxypeptidase B.

for different EC peptides. All three gene products contain several sequences, each of which is processed out to yield physiologically active peptides. If one were to believe all the published reports that have appeared

TABLE III Tissue Distribution Of The Three Enkephalin Genes As Determined By Chemical As Well As Immunological Identification Of The Products

Gene I (Pro-opiocortin)	Gene II (Proenkephalin)	Gene III (Pronorphin)
Ant. Pituitary	Post. Pituitary	Post. Pituitary
Pars Intermedia	Brain	Hypothalamus
	Spinal Cord	Spinal Cord
	Sympathetic Nerves	Intestine
	Intestine	
	Pancreas	
	Adrenal Gland	

products of all three genes are present in almost all mammalian tissues. However, most such reports are based on immunological procedures with no chemical confirmation (Table III).

There have been reports of enkephalins and EC peptides in almost all lower forms of life. We (26) are in the process of examining extracts of many such organisms, insects, earthworms, tetrahymena etc. We do indeed find that crude extracts of most living organisms contain material that crossreacts with antisera to enkephalins and even with the opiate receptor. However, generally after one run over an appropriate HPLC column most such material separates from enkephalins and after further purification on HPLC the "apparent activities" behave unlike any known opioid peptides. Furthermore, treatment of such crossreacting material(s) with trypsin and carboxypeptidase B does not release enkephalins. All such tissues that we have examined contain less than 200 fmol per gram of the free enkephalins or of [Met]enkephalin or [Leu]enkephalin-containing peptides. Whatever the crossreacting materials in lower life forms are, they do not appear to be related to the products derived from proenkephalin or pronorphin.

ACKNOWLEDGMENTS

The studies reported here were carried out by many colleagues. Contributions of former colleagues are noted in the bibliography. Those who are still making major contributions in my laboratory are: Daniel Kilpatrick, Gideon Fleminger, Richard Howells, Elhanan Ezra, Hans-Werner Lahm, Larry Brink and Louise Gerber.

REFERENCES

1. Rubinstein, M., Stein, S., and Udenfriend, S. (1977) *Proc. Natl. Acad. Sci. USA 74*, 4969-4972.
2. Lewis, R.V., Stein, S., Gerber, L.D., Rubinstein, M., and Udenfriend, S. (1978) *Proc. Nat. Acad. Sci. USA 75*, 4021-4023.
3. Schultzberg, M., Lundberg, J.M., Hökfelt, T., Terenius, L., Brandt, J., Elde, R.P., and Goldstein, M. (1978) *Neuroscience 3*, 1169-1186.
4. Lewis, R.V., Stern, A.S., Kimura, S., Rossier, J., Stein, S., and Udenfriend, S. (1980) *Science 208*, 1459-1461.
5. Stein, S. in "Peptides in Neurobiology" (Gainer, H., ed.) pp. 9-37, Plenum Press, New York (1978).
6. Hunkapiller, M.W., and Hood, L.E. (1980) *Science 207*, 523-525.
7. Jones, B.N., Pääbo, S., and Stein, S. (1981) *J. Liq. Chromatogr. 4*, 565-586.
8. Stern, A.S., Lewis, R.V., Kimura, S., Rossier, J., Gerber, L.D., Brink, L., Stein, S., and Udenfriend, S. (1979) *Proc. Nat. Acad. Sci. USA 76*, 6680-6683.
9. Jones, B.N., Shively, J.E., Kilpatrick, D.L., Kojima, K., and Udenfriend, S. (1982) *Proc. Nat. Acad. Sci USA 79*, 1313-1315.
10. Kilpatrick, D.L., Jones, B.N., Kojima, K., and Udenfriend, S. (1981) *Biochem. Biophys. Res. Commun. 103*, 698-705.
11. Stern, A.S., Jones, B.N., Shively, J.E., Stein, S., and Udenfriend, S. (1981) *Proc. Nat. Acad. Sci. USA 78*, 1962-1966.
12. Mizuno, K., Minamino, N., Kangawa, K., and Matsuo, H. (1980) *Biochem. Biophys. Res. Commun. 97*, 1283-1290.
13. Gubler, U., Kilpatrick, D.L., Seeburg, P.H., Gage, L.P., and Udenfriend, S. (1981) *Proc. Nat. Acad. Sci. USA 78*, 5484-5487.

14. Gubler, U., Seeburg, P.H., Gage, L.P., and Udenfriend, S. (1982) *Nature (London) 295*, 206-208.
15. Noda, M., Furutani, Y., Takahashi, H., Toyosato, M., Hirose, T., Inayama, S., Nakanishi, S., and Numa, S. (1982) *Nature (London) 295*, 202-206.
16. Wilson, S.P., Klein, R.L., Chang, K.-J., Gasparis, M.S., Viveros, O.H., and Yang, W.-H. (1980) *Nature (London) 288*, 707-709.
17. Kilpatrick, D.L., Taniguchi, T., Jones, B.N., Stern, A.S., Shively, J.E., Hullihan, J., Kimura, S., Stein, S., and Udenfriend, S. (1981) *Proc. Nat. Acad. Sci. USA 78*, 3265-3268.
18. Goldstein, A., Fischli, W., Lowney, L.I., Hunkapillar, M., and Hood, L. (1981) *Proc. Nat. Acad. Sci USA 78*, 7219-7223.
19. Kangawa, K., Minamino, N., Chino, N., Sakakibara, S., and Matsuo, H. (1981) *Biochem. Biophys. Res. Commun. 99*, 871-878.
20. Kilpatrick, D.L., Wahlström, A., Lahm, H.-W., Blacher, R., Ezra, E., Fleminger, G., and Udenfriend, S. (1982) *Life Sci. 31*, 1849-1852.
21. Fischli, W., Goldstein, A., Hunkapillar, M.W., and Hood, L.E. (1982) *Proc. Nat. Acad. Sci USA 79*, 5435-5437.
22. Kojima, K., Kilpatrick, D.L., Stern, A.S., Jones, B.N. and Udenfriend, S. (1982) *Arch. Biochem. Biophys. 215*, 638-643.
23. Li, C.H. (1977) *Arch. Biochem. Biophys. 183*, 592-604.
24. Lewis, R.V., Stern, A.S., Kilpatrick, D.L., Gerber, L.D., Rossier, J., Stein, S., and Udenfriend, S. (1981) *J. Neurosci 1*, 80-82.
25. Fleminger, G., Ezra, E., Kilpatrick, D.L., and Udenfriend S. Proc. Nat. Acad. Sci. USA (in press).
26. Kilpatrick, D.L. et. al. (to be published).

EXPRESSION OF THE OPIOID GENES[1]

Olivier Civelli
Edward Oates
Haim Rosen
Gerard Martens
Michael Comb
James Douglass
Edward Herbert

Department of Chemistry
University of Oregon
Eugene, Oregon

Opioid peptides are small peptides which exert a morphine-like activity when injected into animals. These peptides contain a met- or leu-enkephalin sequence at their amino terminus and are synthesized in the form of large polypeptide precursors.

In this report, we show that all the opioid peptides described thus far are derived from three different precursors: pro-opiomelanocortin (POMC), proenkephalin and prodynorphin. We show that there is a high level of homology in the organization of these precursors and their corresponding genes in mammalian cells. We also extend our sequence analysis to the structure of the Xenopus laevis proenkephalin gene and how the major exon of this gene resembles its human counterpart. It contains 7 enkephalin sequences, similarly located with respect to the human gene. It presents, however, a difference: the leu-enkephalin sequence of the human gene is replaced by met-enkephalin. These structural analyses indicate that the opioid genes are part of a family of genes which might have evolved from a common ancestral sequence.

[1]*This work was supported by National Institutes of Health Grants AM 16879, AM 30155, and DA 02736.*

ISBN 0-12-417320-9

Knowing the primary sequence of the human proenkephalin gene, we have analyzed this gene for structural differences in different tissues. We have been able to show that this gene indeed exhibits tissue-specific methylation patterns. The human proenkephalin gene in the adrenal gland, for example, is less methylated than it is in the pituitary. The methylation differences are detected essentially between the first and the second exon and in and downstream of the last exon. We then analyzed the expression of the POMC gene in different tissues. We first showed that POMC mRNA is detected in different parts of the brain including the hypothalamus, the amygdala and the cortex. POMC mRNA levels found in these tissues correlate with the POMC-derived peptide levels, indicating that POMC synthesis occurs in the brain regions described. Finally, we have conducted a detailed analysis of the POMC gene expression in the rat pituitary. By sequence analysis we have detected two different POMC mRNAs in this tissue and have demonstrated that these mRNAs differ by 30 bases. These 30 bases are found in the 5' untranslated region of one of the POMC mRNA at the junction between the first and second exon of the gene. The sequence of these 30 bases is identical to the sequence of the first bases in the intron upstream of the second exon. We, therefore, suggest that they are the result of an alternative splicing event. It is important to note that these two forms of POMC mRNA have also been detected in mouse, but not porcine pituitary, indicating that this event is species-specific.

I. INTRODUCTION

In the past 5 years more than 10 peptides have been isolated that exhibit morphinomimetic activity when injected into animals. These peptides, referred to as opioid peptides, have in common an amino terminal met- or leu-enkephalin sequence which is required for biological activity. The opioid peptides which are small in size (ranging from 5 to 31 amino acid residues in length), are synthesized from much larger precursor proteins. This was first demonstrated for β-endorphin, a 31 amino acid opioid, which is synthesized in a precursor which also contains other biologically active peptides such as ACTH and lipotropin (Roberts and Herbert, 1977). Since that discovery, a number of other small neuropeptides have been shown to be synthesized in the form of large precursor molecules from which more than one active peptide can be derived (polyprotein) (Lund *et al.*, 1982;

Land *et al.*, 1982; Goodman *et al.*, 1980).

In an attempt to provide a foundation for analyzing regulation of opioid gene expression, we have determined the structures of the precursors and the genes for the opioid peptides. Using recombinant DNA technology we, and Numa and co-workers, have shown that the synthesis of all the opioid peptides described thus far is controlled by three different genes: one codes for pro-opiomelanocortin (POMC), precursor to β-endorphin, ACTH, β-lipotropin and the melanocyte stimulating hormones (MSH); the second gene codes for proenkephalin which contains 6 met-enkephalin sequences and 1 leu-enkephalin sequence, and gives rise to a variety of opioid peptides in the adrenal medulla including peptides E and F (Udenfriend and Kilpatrick, 1983); the third gene codes for prodynorphin (Proenkephalin B), which contains 3 leu-enkephalin sequences and is the precursor of dynorphin (Goldstein *et al.*, 1979), β-neo-endorphin (Minamino *et al.*, 1981) and rimorphin Kilpatrick *et al.*, 1982).

In this paper we show that the three opioid peptide precursors and the genes that code for these precursors have important structural similarities suggesting a common evolutionary origin. A comparative study of the human and amphibian pro-enkephalin genes is also presented to show how enkephalin peptide structure has evolved in the past 350 million years or so. Finally, we show that diversity in the expression of the POMC gene can be generated at the level of POMC mRNA processing by alternative modes of splicing of this RNA.

II. STRUCTURE OF THE OPIOID PEPTIDE PRECURSORS AND GENES

The three opioid precursors POMC (Nakanishi *et al.*, 1979), proenkephalin (Comb *et al.*, 1982; Noda *et al.*, 1982; Gubler *et al.*, 1982), and prodynorphin (Kakidani *et al.*, 1982) exhibit striking structural similarities (Fig. 1). These precursors are about the same length (257-265 amino acids) and have a similar organization. The bioactive domains are flanked by pairs of basic amino acid residues and are confined to the C-terminal half of the precursor. The amino terminus of the precursor is rich in cysteines and is preceded by a signal peptide that is common to all secreted proteins. Interesting homologies can be seen by comparing the sequences of the proenkephalin and prodynorphin mRNAs.

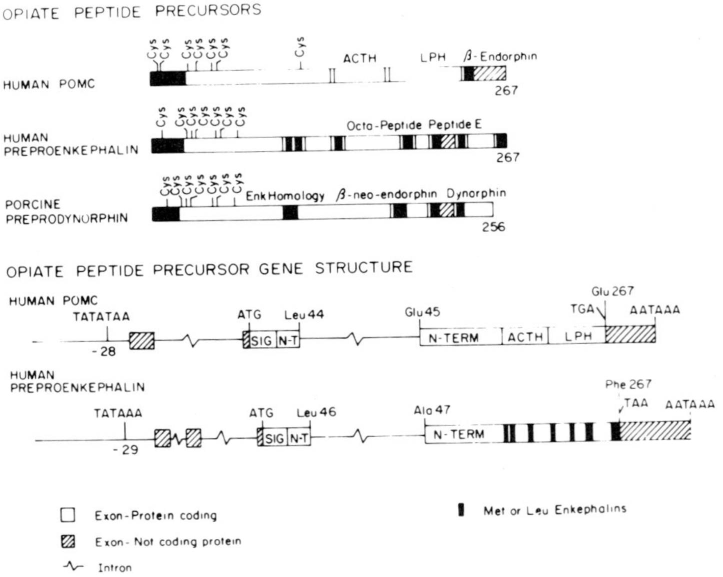

Fig. 1. Schematic representation of the three opioid precursors and of two opioid genes. In the opioid peptide precursors, the black box at the N-terminus represents the signal peptide. The enkephalin homology sequence in the preprodynorphin precursors indicate a sequence found highly homologous to the two first enkephalin units of the preproenkephalin sequence. This enkephalin homology sequence, however, does not code for an enkephalin peptide. In the gene structures SIG and N-T or N-TERM represent respectively the signal peptide and the N-terminal part of the precursors.

First, the location and spacing of the nucleotide sequences that code for the enkephalin units of prodynorphin closely resemble those of the bioactive octapeptide met-enkephalin-Arg-Gly-Leu and of peptide E of proenkephalin. Second, a 124 nucleotide sequence encompassing peptide E exhibits a high degree of homology with a sequence that codes for dynorphin 1-17. Finally a nucleotide sequence encoding units (amino acids 97-105) shows significant homology with the nucleotide sequence encoding the first two met-enkephalin units of proenkephalin. The similarities found at the peptide and the nucleic acid levels are also reflected at the level of gene organization.

Human POMC (Cochet *et al.*, 1982) and proenkephalin genes (Noda *et al.*, 1982; Comb *et al.*, 1983) have been sequenced.

At the 3' end of the genes is a large exon which contains all the bioactive domains and the majority of the N-terminal portions of these precursors. Some 3000 bases upstream (on the 5'-end side), is a smaller exon which contains the rest of the aminoterminus of the precursor, the initiator methionine and a few bases of the 5' untranslated part of the mRNAs. The remainder of the untranslated region of the mRNA is found further upstream in one (POMC) or two (proenkephalin) small exons. Hence, it is clear that the opioid peptide genes belong to a closely related family of genes. The existence of repeated nucleotide sequences in the mRNA coding for the enkephalin units, in the precursors of proenkephalin and prodynorphin indicate that duplication of this unit may account for the origin of opioid peptide genes in the course of evolution.

In this respect, analysis of the opioid genes in different organisms is of interest. The POMC gene in human (Cochet *et al.*, 1982), rat (Drouin and Goodman, 1980), cow (Nakanishi *et al.*, 1981) and mouse (Uhler and Herbert, 1983) shows the same structural arrangement. We have recently analyzed the structure of the proenkephalin gene in *Xenopus laevis*, a toad which diverged from mammals 350 x 10^6 years ago (Fig. 2). The major exon of the gene shows a

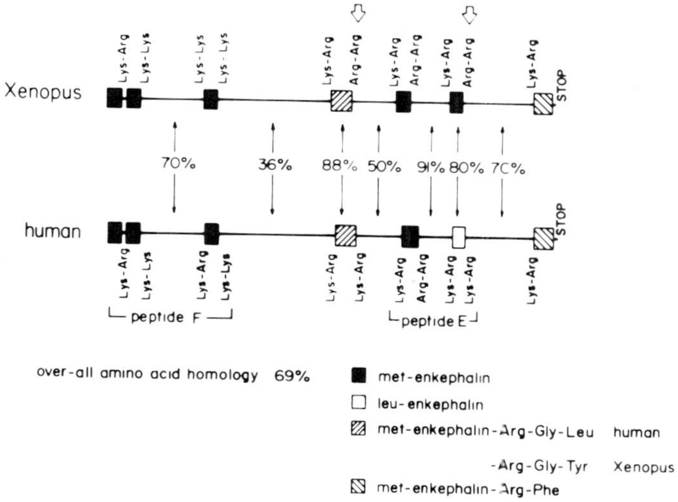

Fig. 2. Sequence comparison of the proenkephalin precursors of human and Xenopus laevis. The Xenopus peptide sequence has been derived from the nucleic acid sequence of a gene isolated from a λ-Xenopus genomic library. The numbers represent the percent of amino acid homology of different stretch of the precursors between the two species.

high degree of homology with its mammalian counterpart at the protein and nucleic acid levels. The *Xenopus* gene contains 7 enkephalin sequences, each flanked by a pair of basic amino acid residues (which suggests for proteolytic cleavage a mechanism similar to that in mammals), and separated by spacer regions of about the same length as those in the human proenkephalin gene. One important difference, however, has been revealed. The amphibian proenkephalin precursor does not contain a leu-enkephalin sequence at the C terminal of peptide E but rather a met-enkephalin sequence. However, peptide E and its amphibian counterpart show a high level of nucleic acid homology suggesting that this peptide has an important biological activity in both species. The similarities in the sequences of the amphibian and human proenkephalin genes show that the basic pattern of the proenkephalin molecule with its 7 enkephalin sequences evolved fairly early in evolution and that the divergence of one of the met-enkephalin sequences to a leu-enkephalin sequence is a recent event.

III. METHYLATION PATTERN OF THE PROENKEPHALIN GENE IN DIFFERENT TISSUES

After determining the sequences of the opioid genes, we turned our attention to factors which affect the expression of these genes. We focused our approach on the methylation patterns of the proenkephalin gene in different tissues.

Changes in DNA methylation patterns have been related to gene expression during differentiation in that the higher the level of expression of a gene the lower the level of methylation of cytosine residues in the dinucleotide sequence, C_pG (Naveh-Many and Cedar, 1981; Felsenfeld and McGhee, 1982). Methylation patterns can be analyzed in a gene by using methylation-sensitive restriction enzymes. The action of some restriction enzymes is prevented by methylation of the cytosine residue in C_pG. The failure of such an enzyme to cut a particular site indicates the presence of a methylated cytosine at that site, thus leading to a direct correlation between the ability of the enzyme to cut a fragment of DNA and the methylation level of the fragment.

DNA from different human tissues was cleaved with three restriction enzymes: SmaI, HpaII and HhaI, transferred to nitrocellulose paper and analyzed for methylation sites in the proenkephalin gene by hybridization with two proenkephalin gene probes. One probe was specific for the 5' end of the gene and one for the large exon of the gene (Comb *et al.*, 1983). The results of these experiments are summarized in Figure 3.

The analysis shows that the 5' end of the proenkephalin gene upstream of Exon I exhibits a low level of methylation for all tissues, but that tissue-specific methylation occurs between Exon I and Exon III. In Exon III and in the first hundred bases downstream from this exon, the proenkephalin gene is highly methylated. In Exon IV, the major exon, and at the 3' end of the gene, tissue-specific differences in methylation are also detected.

A closer look at these data shows that the different tissues analyzed fall into 3 classes. First, adrenal,

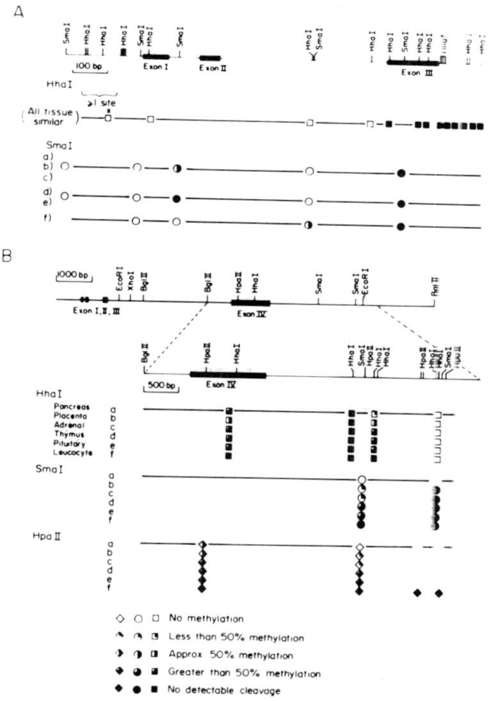

Fig. 3. Methylation patterns of the proenkephalin gene in different human tissues. (A) Methylation surrounding Exons I, II and III. All tissues exhibit a similar degree of methylation at HhaI sites in this region and have been grouped together. Tissues showing similar extent of methylation at SmaI sites have been grouped together.
(B) Methylation surrounding Exon IV. Evaluation of the levels of methylation is indicated at each site for each tissue. a) pancreas; b) placenta; c) adrenal; d) thymus; e) pituitary; f) leucocyte.

pancreas and placenta are partly methylated between Exon I and II and more undermethylated than the other tissues at their 3' end. Thymus and pituitary, which are highly methylated between Exon I and II, have a high level of methylation at their 3' end. Finally, the proenkephalin gene in leucocytes is not methylated at its 5' end and highly methylated at its 3' end. It is interesting to note that methylation specific classes can be related to the levels of the proenkephalin gene expression in these tissues since adrenal expresses proenkephalin mRNA and pancreas and placenta contains proenkephalin peptides, whereas pituitary is not expressing proenkephalin (Comb *et al.*, 1983).

These studies show structural differences of the proenkephalin gene in different tissues. They represent therefore, an important basis for our studies on structural features involved in mechanisms underlying the expression of the proenkephalin gene. It is now possible to carry out *in vitro* methylation of the proenkephalin gene at the same sites as described above and analyze the effect of specific methylation on the level of gene expression by gene-transfer experiments.

IV. TISSUE-SPECIFIC EXPRESSION OF THE OPIOID GENES

Opioid peptides are found in a wide variety of tissues. POMC-derived peptides are present in the pituitary, in different parts of the brain, in the placenta and in the intestine. Peptides derived from proenkephalin have also been detected in some of these tissues but also occur in the sympathetic ganglia, the pancreas, and in the adrenal medulla. Prodynorphin peptides have a significant overlap in their distribution with proenkephalin peptides but occur also in different cells (Höllt, 1983). To determine the exact sites of synthesis of these peptides we have examine the distribution of mRNA that codes for the opioid peptides in different tissues. First, we analyzed mRNA (poly(A) RNA) from pituitary and different parts of the rat brain by the Northern blot procedure and solution hybridization assay to determine the content of POMC mRNA in each tissue. We used a cloned mouse POMC cDNA as the hybridization probe in these assays (Fig. 4). These results showed that POMC mRNA is present in high concentrations in the two lobes of the pituitary, and that POMC sequence can also be detected in much lower concentrations in the hypothalamus, amygdala and cerebral cortex (Civelli *et al.*, 1982). The POMC mRNA levels determined in these tissues correlate well with the

the peptide levels and indicate that POMC gene transcription is not restricted to the pituitary but also occurs in the brain.

V. DIFFERENTIAL EXPRESSION OF THE POMC GENE IN THE RAT PITUITARY

Several reports have indicated that several forms of POMC are synthesized in rat pituitary (Oates and Herbert, 1983). By analysis of the products obtained by cell-free translation of POMC mRNA, we have been able to determine that the heterogeneity resides in the N-terminal region of the precursor (Oates, E., personal communication). We decided therefore, to analyze the sequences in the 5' region of POMC mRNA in

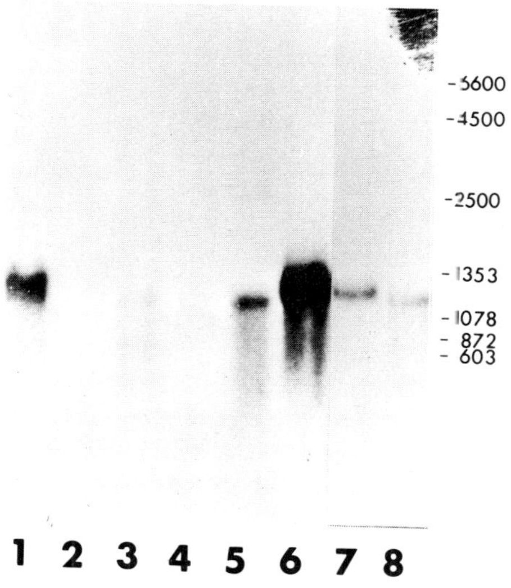

Fig. 4. POMC mRNA in rat brain tissues. Poly (A) mRNA was extracted from different tissues, subjected to electrophoresis and transferred to nitrocellulose paper. Hybridization was done with a mouse DNA fragment specific to the LPH region of the POMC precursor. Lane 1: hypothalamic mRNA, 9 μg; Lane 2: cerebellum mRNA, 25 μg; Lane 3: cortex mRNA, 25 μg; Lane 4: midbrain mRNA, 25 μg; Lane 5: amygdala mRNA, 25 μg; Lane 6: AtT20 mRNA, 0.07 μg; Lane 7: neurointermediate pituitary mRNA, 0.009 μg; Lane 8: amygdala mRNA, 25 μg (different preparation).

rat pituitary. We used a specific cDNA priming technique to determine these sequences. Two synthetic oligonucleotides were synthesized complementary to two different regions of the 5' end of the mRNA. Each oligmer was labelled with ^{32}P at its 5' end and used as a primer for the synthesis of cDNA with reverse transcriptase. The cDNA products were separated by gel electrophoresis and sequenced (Oates and Herbert, 1983). Figure 5 shows that rat mRNA serves as template for the synthesis of two different cDNA products. These products differ in that one of them has a 30 base insert in the 5' untranslated region of the mRNA at the exact site of the junction between the first and the second exon of the gene. Moreover this 30 base insert has the same sequence as the first 30 bases upstream of the second exon. Hence, at least two forms of POMC mRNA are synthesized in rat pituitary. These forms of POMC mRNA are present in both the anterior and intermediate lobes of rat pituitary in the ratio of approximately 5 to 1 (amount of form without the insert to amount of form with the insert). In addition both forms of the POMC mRNA are found in the cytoplasm of the pituitary cells. Since it appears likely that there is only one POMC gene in the rat (Drouin, personal communication), these results indicate that differential expression of the POMC gene in the rat pituitary could be the result of an alternative RNA splicing event. The implications of this discovery are not yet understood. Expression of two or more mRNA species from a single gene could be of great importance in the post-transcriptional regulation of a gene. It is also important to note that the two types of POMC mRNA found in rat pituitary are also found in mouse pituitary, but not in porcine pituitary, indicating that the alternate splicing event is species-specific.

VI. CONCLUDING REMARKS

These data show that the three different opioid peptide precursors are part of the growing family of polyproteins and that they share striking homologies at the peptide and nucleic acid levels. The similarities detected between the proenkephalin and prodynorphin genes suggest that these genes may have evolved from a common ancestral gene. In this respect, our analysis of the sequence of the amphibian proenkephalin gene is revealing. This particular gene compared to its human counterpart shows a high level of homology in the biologically active domains and divergence in the spacer regions. The amphibian gene contains only met-enkephalin sequences indicating that the presence of a leu-enkephalin in the human or the bovine proenkephalin genes is a relatively early event. The degree of methylation of proenkephalin genes in different tissues does not relate directly to tissue-specific levels of expression of this gene but indicates significant structural differences of the

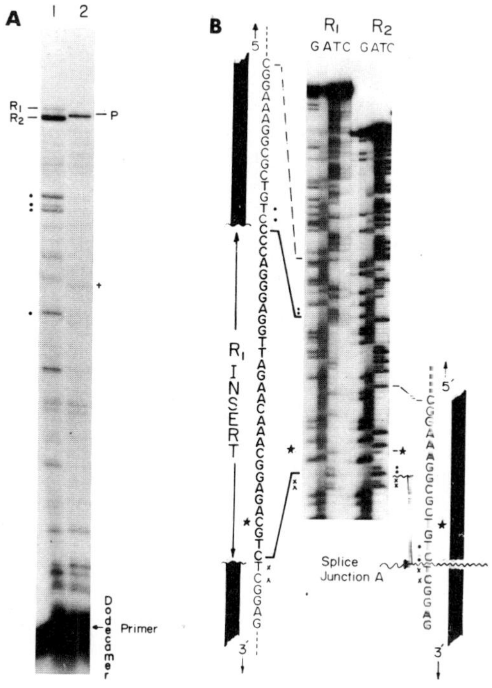

Fig. 5. Sequence of the two rat pituitary POMC mRNA. Poly(A) mRNA isolated from rat pituitary was hybridized with a synthetic radioactively labelled oligodeoxynucleotide probe complementary to the amino acids 57 - 60 of the precursor. Primer extension was achieved by the action of the reverse transcriptase in presence of non-labelled nucleotides. The products of this reaction were subjected to electrophoresis on 8% urea-polyacylamide gel (A). The two bands R_1 and R_2 were further sequenced by the Maxam and Gilbert technique (B). In (A), Lane 1 shows the results of rat pituitary mRNA primary; Lane 2, a similar result obtained with porcine pituitary mRNA. The dots represent nonterminated POMC products; the cross a non-POMC-related product. In (B), the sequences of the two rat products is shown. A 30 bases addition is found in R_1, inserted at splice junction A which indicates the junction site between Exon I and Exon II in the rat gene.

proenkephalin gene in different human tissues. In our tissue survey of POMC gene expression we are able to show that in the hypothalamus, the amygdala and the cerebral cortex, POMC mRNA is present at a level corresponding to the POMC peptide level. Therefore, these peptides are very likely synthesized in the sites in the brain where they are detected and not transported to these sites from the pituitary.

Finally, by using specific cDNA priming reactions we show that the rat pituitary POMC gene is transcribed in the form of two different mRNAs which differ in their 5' untranslated region. This result which can be attributed to differential splicing events, is species-specific and could have implications in the post-transcriptional regulation of the rat pituitary POMC gene. Our studies, therefore, open the way to a more thorough understanding of the mechanism underlying the regulation of the opioid peptide gene family.

REFERENCES

Civelli, O., Birnberg, N. and Herbert, E. (1982). *J. Biol. Chem. 257*, 6783-6784.

Cochet, M., Chang, A.C.Y. and Cohen, S. (1982). Nature *297*, 335-339.

Comb, M., Seeburg, P., Adelman, J., Eiden, L. and Herbert, E. (1982) Nature *295*, 663-666.

Comb, M., Rosen, H., Seeburg, P., Adelman, J. and Herbert, E. (1983) DNA (submitted for publication).

Drouin, J. and Goodman, H.M. (1980). Nature *288*, 610-613.

Felsenfeld, G. and McGhee J. (1982). Nature *296*, 602-603.

Goldstein, A., Tachibana, S., Lowney, L.I., Hunkapiller, M. and Hood, L. (1979). *Proc. Natl. Acad. Sci. USA 76*, 6666-6670.

Goodman, R.H., Jacobs, J.W., Chin, W.W., Lund, P.K., and Dee, P.C. (1980). *Proc. Natl. Acad. Sci. USA 77*, 5869-5873.

Gubler, U., Seeburg, P., Hoffman, B.J., Gage, L.P. and Udenfriend, S. (1982). Nature *295*, 206-208.

Höllt, V. (1983). *TINS 6*, 24-26.

Kakidani, H., Furutani, Y., Takahashi, H., Noda, M., Morimoto, Y., Hirose, T., Asai, M., Inayama, S., Nakanishi, S. and Numa, S. (1982). Nature *298*, 245-249.

Kilpatrick, D.L., Jones, B.N., Lewis, R.V., Stern, A.S., Kojima, K., Shively, J.E. and Udenfriend, S. (1982). *Proc. Natl. Acad. Sci. USA 79*, 3057-3061.

Land, H., Schütz, G., Schmale, H. and Richter, D. (1982). Nature *295*, 299-303.

Lund, P.K., Goodman, R.H., Dee, P.C. and Habener, J.F. (1982). *Proc. Natl. Acad. Sci. USA 79*, 345-349.

Minamino, N., Kangawa, K., Chino, N., Sakakibara, S. and Matsuo, H. (1981). *Biochem. Biophys. Res. Comm. 99*, 864-870.

Nakanishi, S., Inoue, A., Kita, T., Nakamura, M., Chang, A.C.Y., Cohen, S. and Huma, S. (1979). Nature *278*, 423-427.

Nakanishi, S., Teranishi, Y., Watanbe, Y., Notake, M., Noda, M., Kakidani, H., Tingami, H. and Numa, S. (1981). *Eur. J. Biochem. 115*, 429-438.

Naveh-Many, T. and Cedar, H. (1981). *Proc. Natl. Acad. Sci. USA 78*, 4246-4250.

Noda, M., Furutani, H., Takahashi, H., Toyosato, M., Hirose, S., Inayama, S., Nakanishi, S. and Numa, S. (1982). Nature *295*, 202-206.

Noda, M., Teranishi, Y., Takahashi, H., Toyosato, M., Notake, M., Nakanishi, S. and Numa, S. (1982). Nature *297*, 431-434.

Oates, E. and Herbert, E. (1983). Nature (submitted for publication).

Roberts, J. and Herbert, E. (1977). *Proc. Natl. Acad. Sci. USA 74*, 5300-5304.

Udenfriend, S. and Kilpatrick, D.L. (1983). *Arch. Biochem. Biophys. 221*, 309-323.

Uhler, M., Herbert, E., D'Eustachio, P. and Ruddle, F.D. (1983). *J. Biol. Chem.* (in press).

RECEPTOR SELECTIVITIES OF THE THREE OPIOID PEPTIDE FAMILIES

Volker Höllt and Bernd R. Seizinger

Department of Neuropharmacology
Max-Planck-Institut für Psychiatrie
München, F.R.G.

Javier Garzón and Horace H. Loh

Department of Pharmacology
University of California
San Francisco, CA

A large number of opioid peptides deriving from three distinct precursor molecules (pro-opiomelanocortin = POMC, pro-enkephalin A, pro-enkephalin B) were tested for their opioid activity in various test systems to characterize the selectivity of the three peptide families for the three pharmacologically distinguishable opioid receptor subtypes mu, kappa, delta.

ß-Endorphin - endproduct in the processing of POMC - shows selectivity for mu and delta, but not for kappa opioid receptors.

Pro-enkephalin B generates a variety of opioid peptides with a high selectivity for kappa (dynorphins, neoendorphins) and delta receptors ([Leu^5]-enkephalin) but not for mu receptors.

Pro-enkephalin A gives rise to peptides with selectivity for all three receptor subtypes: mu (BAM-22P), delta (enkephalins) and kappa ([Met^5]-enkephalin-Arg^6-Gly^7-Leu^8).

After intracerebroventricular injection into the mouse brain, peptides with preference for mu opioid receptors (ß-endorphin and the majority of pro-enkephalin A derived peptides) induce analgesia supporting the hypothesis that antinociceptive action of opioids might be predominantly mediated via mu receptors.

BIOCHEMICAL AND CLINICAL ASPECTS OF NEUROPEPTIDES: SYNTHESIS, PROCESSING, AND GENE STRUCTURE

ISBN 0-12-417320-9

INTRODUCTION

Numerous opioid peptides have been isolated which contain either the sequence of [Met^5]-enkephalin or [Leu^5]-enkephalin at the N-terminal. All these peptides belong to three peptide families each deriving from a distinct precursor molecule (for a review, see Höllt, 1983).

Besides pro-opiomelanocortin (POMC) - the common precursor for ACTH and ß-endorphin (Nakanishi et al., 1979) - two precursors for the enkephalins have recently been analyzed using recombinant DNA techniques (Noda et al., 1982; Gubler et al., 1982; Kakidani et al., 1982).

Pro-enkephalin A (also termed pro-enkephalin (Gubler et al., 1982)) contains four copies of [Met^5]-enkephalin, one of [Leu^5]-enkephalin, one of the hepta-peptide [Met^5]-enkephalin-Arg^6-Phe^7 and one of the octa-peptide [Met^5]-enkephalin-Arg^6-Gly^7-Leu^8 (Noda et al., 1982; Gubler et al., 1982) each peptide flanked by pairs of basic aminoacids. In addition, several larger pro-enkephalin A-derived opioid peptides (see Figure 2) have been isolated from bovine adrenal medulla, such as peptide E, peptide F (Kilpatrick et al., 1981), BAM-12P (Mizuno et al., 1980a), BAM-20P, BAM-22P (Mizuno et al., 1980b).

Pro-enkephalin B (also termed pro-dynorphin by Goldstein's group (Fischli et al., 1982)) contains three copies of [Leu^5]-enkephalin flanked by pairs of basic aminoacids (Kakidani et al., 1982). Furthermore, it contains a series of larger opioid peptides, which has been isolated from porcine pituitary and brain, such as α-neoendorphin (Kangawa et al., 1981), ß-neoendorphin (Minamino et al., 1981), dynorphin A (Fischli et al., 1982), $dynorphin_{1-8}$ (Minamino et al., 1980), dynorphin B (=rimorphin, Kilpatrick et al., 1982), dynorphin-32 - a common precursor for dynorphin A and dynorphin B (Fischli et al., 1982). Finally, the existence of a larger dynorphin B containing peptide (here termed dynorphin B-29) has been proposed on the basis of the mRNA structure (Kakidani et al., 1982).

It was the aim of the present study to characterize the interaction of these numerous peptides with their corresponding receptors. Pharmacological evidence for the multiplicity of opioid receptors has been accumulated during the past years (Martin et al. 1976; Lord et al., 1977; for a review see Wüster et al., 1981). There is now general agreement that opioi receptors can be classified into at least three subtypes: i) the 'mu' receptor for which morphine is a prototype ligand; ii) the 'kappa' receptor for which ethylketacyclazocine is a prototype ligand; iii) the 'delta' receptor for which the enkephalins are prototype ligands. It will be shown that there is a certain degree of preference of peptides encoded by the different genes to these receptors: POMC generates ß-endorphin - a peptide with selectivity for mu and delta opioid receptors but not for kappa receptors; peptides generated from pro-enkephalin B occupy preferentially kappa and delta but not mu receptors; the pro-enkephalin A gene shows the least selectivity; it encodes peptides which have preferential affinity either for mu or delta or kappa opioid receptors.

MATERIALS AND METHODS

Analgesia was measured by the radiant heat tail flick method after intracerebroventricular injection of the opioids into male ICR mice (22-25 g) as described (Höllt et al., 1982b).

For the rabbit vas deferens bioassay male New Zealand white rabbits (2500-3000 g) were decapitated, the prostatic part of the vasa mounted in 10 ml tissue chambers containing Krebs solution as described (Oka et al., 1980). A basal tension of 500-700 mg was applied and the tissues electrically stimulated by supramaximal rectangular pulses (0.1 Hz; 1 msec duration). To minimize degradation of the peptides enzyme inhibitors were added to the bath fluid (bestatin (30 μM), captopril (10 μM) and l-leucyl-l-leucine (2 mM)) as proposed by McKnight et al., 1983). The inhibitors had no effect on the response of the vasa to field stimulation. Usually three concentrations of the peptides were tested to construct a dose-response curve (Rezvani et al., 1983).

For binding experiments, membranes were prepared from brains of male ICR mice (20-25 g). The membranes diluted with HEPES buffer (25 mM; pH 7.7) to a final protein concentration of 0.6 to 0.7 mg/ml were incubated in the presence of tritiated ligands and unlabelled peptides for 180 min at 0°C. ^{3}H-dihydromorphine (DHM) and ^{3}H-D-Ala^{2}-D-Leu^{5}-enkephalin (DADLE) were used at 0.5 nM and ^{3}H-ethylketacyclazocine (EKC) at 3 nM final concentrations; in the last case, the assay mixture also contained 100 nM FK 33824 (Sandoz) and 100 nM unlabelled DADLE to block the access of ^{3}H-EKC to mu and delta receptor sites respectively (Kosterlitz et al., 1981). Nonspecific binding was determined in the presence of 10 μM naloxone. Free was separated from bound ligands by filtration of the mixture (final volume 2 ml) through glass fiber filters (Whatman GF/B).

In separate experiments the dissociation constants (K_{Ds}) of the protype ligands to their respective receptor sites were determined: DHM to mu sites: 1.6 ± 17 nM; DADLE to delta sites: 1.5 ± 0.15 nM; EKC to kappa site 0.2 ± 0.05 nM (mean ± SEM; n = 3). The K_{Ds} were used to calculate the inhibition constants (K_{I}s) for each peptide (see text).

All peptides were purchased from Peninsula Labs (San Carlos, CA, USA).

RESULT AND DISCUSSION

Recently, we reported that several opioid peptides can induce analgesia after i.c.v. injection into the brain of mice (Höllt et al., 1982b). Table I summarizes these results. It clearly illustrates that besides ß-endorphin - the most potent opioid peptide - the majority of peptides derived from pro-enkephalin A produce analgesia, whereas the enkephalins and all peptides derived from pro-enkephalin B are inactive. The most potent pro-enkephalin A derived peptide BAM-22P (Mizuno et al., 1980b) has an about 6-fold greater analgesic potency than morphine on a molar basis. This finding indicates that the pro-enkephalin B-derived peptides might interact with opioid receptors different from that which might mediate analgesia in the brain.

To characterize the receptor selectivities of the three opioid peptide families, bioassay and binding experiments were performed. A particularly useful bioassay is the isolated vas deferens preparation of the rabbit, since the electrically induced contractions of this tissue can only be inhibited by opioids which are able to interact with kappa receptors (Oka et al., 1980).

Figure 1 shows the relative potencies of a wide variety of opioid peptides derived from three distinct opioid precursors. The activity is expressed in percent of that of dynorphin A, the most potent opioid peptide. The potency of the peptides has been determined in the presence of enzyme inhibitors. This minimized the degradation of the peptides and markedly increased the potency of the 'smaller' peptides such as $dynorphin_{1-8}$ and ß-neoendorphin

TABLE I.

Opioid	ED_{50} (nmol/mouse) i.c.v.	time of peak effect (min)
<u>Pro-opiomelanocortin</u>		
Human ß-endorphin	0.05 (0.03-0.08)	30
<u>Pro-enkephalin A</u>		
[Leu^5]-enkephalin	inactive	
[Met^5]-enkephalin	inactive	
[Met^5]-enkephalin-Arg^6-Phe^7	49.6 (35.4- 69.4)	≤10
[Met^5]-enkephalin-Arg^6-Gly^7-Leu^8	105.7 (70.5-158.5)	≤10
BAM-12P	56.2 (35.8- 87.0)	≤10
BAM-22P	1.5 (1.0- 2.1)	20
Peptide E	2.1 (1.6- 2.9)	20
Peptide F	3.0 (1.9- 4.6)	20
<u>Pro-enkephalin B</u>		
α-Neoendorphin, ß-neoendorphin	inactive	
dynorphin A, dynorphin A(1-8)	inactive	
dynorphin B, dynorphin B-29	inactive	
dynorphin-32	inactive	
Morphine	8.8 (6.0- 13.0)	30

Antinociceptive effect of endogenous opioid peptides and morphine in the tail flick test after i.c.v. injection into mice.

Values represent means ± 95% confidence limits. ED_{50} at time of peak effects were calculated from dose-response curves constructed from at least 3 doses, each dose given to 8-10 mice. 'Inactive' means ED_{50} greater than 100 nmol/mouse.

(Corbett et al., 1982; Rezvani et al., 1983). Figure 1 shows that (with the exception of [Leu^5]-enkephalin) all pro-enkephalin B-derived peptides

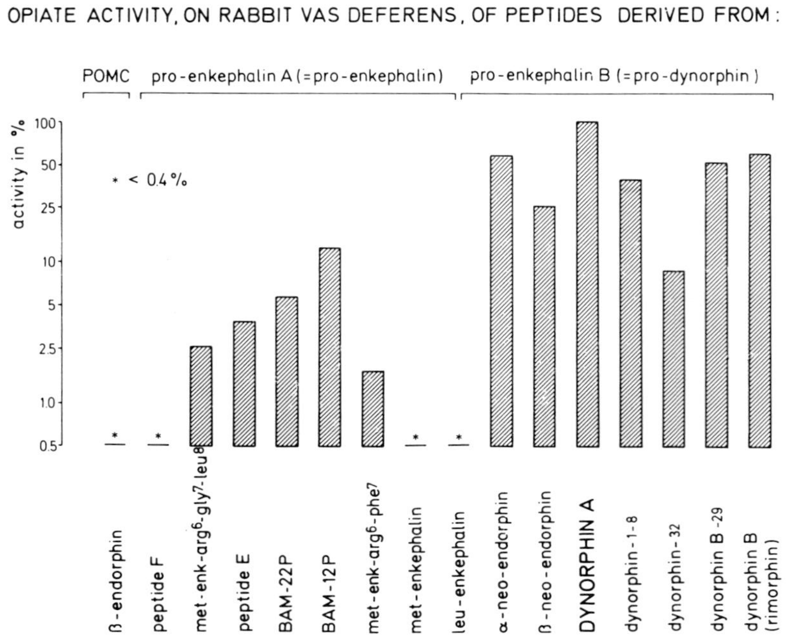

FIGURE 1. Relative potencies of endogenous opioid peptides on the rabbit vas deferens.

The effect of the peptides to inhibit the electrically stimulated vas preparation (IC_{50}) was measured in the presence of enzyme inhibitors (L-leucyl-L-leucine: 2 mM; bestatin: 30 µM; captopril: 10 µM). Bars represent the relative activity in percent of that of dynorphin A (IC_{50} = 218 ± 140 nM; mean ± SEM; n = 4), means of 3 to 5 determinations.

are active on this preparation indicating that all these peptides are able to interact with kappa opioid receptors. In contrast, ß-endorphin - endproduct in the processing of POMC and the enkephalins - putative endproducts in the processing of pro-enkephalin A and pro-enkephalin B - are virtually inactive on this preparation. Moreover, these peptides do not have antagonistic activity on this preparation (data not shown). Thus, these observations indicate that these peptides have no or a very weak affinity for kappa opioid receptors. Interestingly, with the exception of peptide F and the enkephalins, all pro-enkephalin A-derived peptides show a significant activity on this preparation, although less than the pro-enkephalin B-derived peptides, indicating interaction with kappa opiate receptors. The kappa receptor activity appears to increase, if peptide E is gradually shortened at the C-terminus into BAM-22P and BAM-12P.

To further analyze the receptor selectivities of the pro-enkephalin A-derived peptides binding studies were performed. In these experiments the ability of the opioid peptides to displace radioactive prototype ligands from their specific binding sites in mouse brain was measured. ^{3}H-dihydromorphine was used to label mu sites, ^{3}H-D-Ala^2-D-Leu^5-enkephalin to label delta sites and ^{3}H-ethylketacyclazocine to label kappa sites. Table II summarizes the results. The affinities of the opioid peptides are expressed as inhibition constants (K_{I}s) and were calculated according the formula: $K_I = IC_{50}\ (1+L/K_D)$; (where IC_{50} is the concentration of a peptide to displace a given concentration of the radioactive ligand (L) by 50%; K_D is the dissociation constant of the receptor interaction of the radioactive prototype ligand).

Table II shows marked potency differences in the binding to mu and kappa sites between the various pro-enkephalin A-derived peptides. For instance, peptide E and its fragments BAM-22P and BAM-12P have much higher affinities to mu and kappa receptors than peptide F or [Leu^5]-enkephalin. On the other hand, all opioid peptides (with the exception of peptide F) exhibit a similar affinity for delta opioid receptors.

To characterize the receptor selectivities and affinities of the peptides in a more illustrative way two additional binding parameters were calculated according to McKnight et al., 1982: i) the total affinity ($K_{A(t)}$) of a peptide to all three opioid receptor subtypes:

$$K_{A(t)} = 1/K_{I(t)} = 1/K_{I(mu)} + 1/K_{I(delta)} + 1/K_{I(kappa)}$$

ii) the relative affinity (RA) as a measure of the receptor selectivity:

$$RA_{(mu)} = \frac{1/K_{I(mu)}}{1/K_{I(t)}}; \quad RA_{(delta)} = \frac{1/K_{I(delta)}}{1/K_{I(t)}}; \quad RA_{(kappa)} = \frac{1/K_{I(kappa)}}{1/K_{I(t)}}$$

TABLE II.

Peptide	K_I (nM) vs [^{3}H]-DHM (mu)	[^{3}H]-DADLE (delta)	[^{3}H]-EKC (kappa)
[Leu^5]-enkephalin	68.2 ± 10.8	9.4 ± 1.8	3697.0 ± 420.0
[Met^5]-enkephalin	22.3 ± 2.1	5.8 ± 1.1	1566.0 ± 221.0
[Met^5]-enkephalin-Arg^6-Phe^7	6.9 ± 1.0	3.9 ± 1.2	19.2 ± 4.3
[Met^5]-enkephalin-Arg^6-Gly^7-Leu^8	23.7 ± 1.4	9.0 ± 1.0	9.3 ± 2.1
BAM-12P	0.7 ± 0.2	6.0 ± 1.1	1.9 ± 0.8
BAM-22P	0.9 ± 0.2	7.2 ± 2.0	2.9 ± 1.6
Peptide E	2.0 ± 0.5	5.6 ± 1.0	7.3 ± 0.9
Peptide F	178.0 ± 24.8	131.0 ±18.3	82.0 ± 13.4
Dynorphin A	2.6 ± 0.9	6.3 ± 0.9	0.3 ± 0.1
Human ß-endorphin	1.0 ± 0.1	4.0 ± 0.1	15.6 ± 1.5

Affinities of opioid peptides to mu, delta and kappa opioid receptor sites in mouse brain membranes.
The mu-site was labelled with 0.5 nM [^{3}H]-dihydromorphine (DHM); the delta site with 0.5 nM [^{3}H]-D-Ala2-D-Leu^5-enkephalin (DADLE) and the kappa-site with 3 nM [^{3}H]-ethylketocyclazocine (EKC) in the presence of 100 nM FK 33824 (Sandoz) and 100 nM DADLE. For details see Methods.
Affinities are expressed as K_{Is} (inhibition constants) and computed as described in the text. Values are the mean ± SEM of at least three determinations.

Figure 2 shows a schematic diagram of pro-enkephalin A and its putative processing products. The binding characteristics of these opioid peptides to the receptor subtypes in mouse brain are graphically illustrated as circles: The magnitude of the total affinity ($K_{A(t)}$) is proportional to the circle area; the relative affinity (RA) is indicated by the size of the three different segments within the circle. As evident from figure 2, there are marked differences in both the total affinities and the receptor

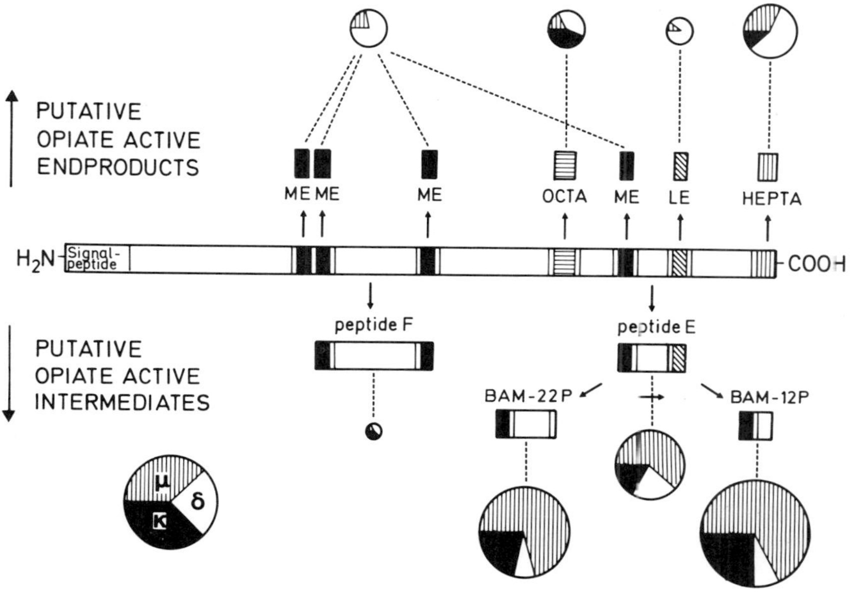

FIGURE 2. Receptor selectivities of opioid peptides derived from pro-enkephalin A.

The total affinities ($K_{A(t)}$) and the relative affinities (RA) for mu, delta and kappa receptor sites have been calculated using the data listed in table II and the formulae described in the text. The figure shows the data in a schematic form: The size of the circle area is proportional to the $K_{A(t)}$ and the size of segments indicate the RAs for mu, delta and kappa sites.

The bar represents bovine pre-pro-enkephalin A and the opioid peptides derived from it (see Noda et al., 1982). The putative processing endproducts ME = [Met^5]-enkephalin, LE = [Leu^5]-enkephalin, HEPTA = [Met^5]-enkephalin-Arg^6-Phe^7 and OCTA = [Met^5]-enkephalin-Arg^6-Gly^7-Leu^8 are symbolized by black or hatched bars. The double basic aminoacids are indicated by the vertical lines flanking these peptides.

selectivities between the pro-enkephalin A-derived peptides. BAM-12P and BAM-22P have the highest, peptide F and [Leu^5]-enkephalin the lowest overall affinity to the three opioid receptor subtypes. BAM-22P is the peptide with the highest selectivity for mu receptors ($RA_{(mu)}$ = 0.70). The octapeptide [Met^5]-enkephalin-Arg^6-Gly^7-Leu^8 has the highest kappa receptor selectivity ($RA_{(kappa)}$ = 0.45) and [Leu^5]-enkephalin is the peptide with the highest preference for delta opioid receptors ($RA_{(delta)}$ = 0.88). An interesting finding is the high mu receptor selectivity of peptide E and its fragments BAM-22P and BAM-12P. In particular, the relative affinity of BAM-22P to mu opioid receptors ($RA_{(mu)}$ = 0.70) is similarly high as that of ß-endorphin ($RA_{(mu)}$ = 0.76; $RA_{(delta)}$ = 0.20; $RA_{(kappa)}$ = 0.04) as calculated from the data in table II.

On the other hand, dynorphin A - processing product of pro-enkephalin B - has a low relative affinity to mu, but a high selectivity for kappa opioid receptors ($RA_{(mu)}$ = 0.10; $RA_{(delta)}$ = 0.04; $RA_{(kappa)}$ = 0.86; calculated from data in table II.

A low relative binding affinity for mu opioid receptors has also been found for other pro-enkephalin B-derived peptides, such as α-neoendorphin, $dynorphin_{1-8}$, dynorphin B (McKnight et al., 1982; Garzón et al., 1983). These observations indicate that peptides with high mu receptor selectivity are generated from POMC (ß-endorphin) and pro-enkephalin A (BAM-22P), but not from pro-enkephalin B. Conversely, the pro-enkephalin B-derived peptides (with the exception of [Leu^5]-enkephalin) exhibit all a high selectivity and affinity for kappa opioid receptors. This has been first demonstrated for the dynorphin A fragment $dynorphin_{1-13}$ (Wüster et al., 1980; Huidobro et al., 1981; Chavkin and Goldstein, 1982) and later for dynorphin A (Wüster et al., 1981; Oka et al., 1982; McKnight et al., 1982; Schulz et al., 1982), $dynorphin_{1-8}$ (McKnight et al., 1982), α-neoendorphin, ß-neoendorphin (Oka et al., 1982; McKnight et al., 1982; Schulz et al., 1982), dynorphin B and dynorphin-32 (Fischli et al., 1982).

In contrast, very few reports exist showing kappa receptor activities of peptides derived from pro-enkephalin A. The heptapeptide [Met^5]-enkephalin-Arg^6-Phe^7 was first reported to be a selective kappa receptor ligand (Audigier et al., 1980). However, it was later found that it exhibits selectivity for delta opioid receptors (Wüster et al., 1981; see also Figure 2). Recently, the kappa receptor activity of the octapeptide [Met^5]-Arg^6-Gly^7-Leu^8 was reported by Schulz et al., 1982. Although the octapeptide is the pro-enkephalin A-derived peptide with the highest selectivity for kappa opioid receptors, the highest kappa receptor affinity resides in peptide E, BAM-22P and BAM-12P, peptides with preference for mu opioid receptors (Figure 1,2, Table II). BAM-12P has a higher kappa receptor selectivity than peptide E its precursor peptide. Thus, a putative processing of peptide E into BAM-22P and BAM-12P would continuously increase the kappa selectivity of the generated peptides. Such a processing, however, would involve unusual processing signals rather than the cleavage at double (or single) basic aminoacids (Figure 2). The existence of substantial amounts of BAM-12P and BAM-22P in bovine adrenal medulla (Mizuno et al., 1980a,b; Höllt et al., 1981) and brain (Baird et al., 1982; Höllt et al., 1982a), however, provide strong evidence for such an 'unusual' peptide processing in these tissues. Although these peptides have a high selectivity for mu opioid receptors, their substantial kappa receptor affinity and their presence in the brain suggest that they might also interact with kappa receptors in the nervous system. In fact, immunohistochemical studies have localized BAM-22P in cells and fibers of cortical structures (Khatchaturian et al., 1983) which have been found to have a very high kappa receptor density (Goodman and Snyder, 1982).

These findings indicate that besides the pro-enkephalin B-derived peptides, some of the pro-enkephalin A-derived are able to interact with kappa opioid receptors with substantial affinity. In contrast, ß-endorphin-processing product of POMC - does not have any appreciable affinity for kappa opioid receptors.

It has been hypothesized that the antinociceptive action of opioids might be mediated by their interaction with mu opioid receptors (Kosterlitz et al., 1980). This assumption appears to be generally supported by our findings. ß-Endorphin and various pro-enkephalin A-derived peptides with high selectivity for mu opioid receptors are analgesically effective, in contrast to the enkephalins and the pro-enkephalin B-derived peptides which have a low preference for mu opioid receptors. Exceptions are the low analgesic potency of BAM-12P and the high potency of peptide F (Table I). Such discrepancies might be caused by the degradation of the peptides in mouse brain in vivo: degradation products might be less (BAM-12P) or even be more potent (peptide F) than the unmetabolized peptides.

REFERENCES

Audigier, Y., Mazarguil, H., Rossier, J. & Cros, J. (1980). Europ. J. Pharmacol. 68, 237-238.
Baird, A., Ling, N., Böhlen, P., Benoit, R., Klepper, R. & Guillemin, R. (1982). Proc. Natl. Acad. Sci. USA 79, 2023-2025.
Chavkin, C., Jones, I.F. & Goldstein, A. (1982). Science 215, 413-415.
Corbett, A.D., Paterson, S.J., McKnight, A.T., Magnan, J. & Kosterlitz, H.W. (1982). Nature 299, 79-81.
Fischli, W., Goldstein, A., Hunkapiller, M.W. & Hood, L.E. (1982). Proc. Natl. Acad. Sci. USA 79, 5435-5437.
Garzón, J., Sánchez-Blazquez, P., Höllt, V., Lee, N.M. & Loh, H.H. (1983). Life Sci., in press.
Goodman, R.R. & Snyder, S.H. (1982). Proc. Natl. Acad. Sci. USA 79, 5703-5707.
Gubler, U., Seeburg, P., Hoffman, B.J., Gage, L.P. & Udenfriend, S. (1982). Nature 295, 206-208.
Höllt, V., Grimm, C., Haarmann, I., Seizinger, B.R. & Herz, A. (1981). In "Advances in Endogenous and Exogenous Opioids", H. Takagi & E.J. Simon (eds.), pp. 152, Elsevier/North-Holland, Amsterdam.
Höllt, V., Haarmann, I., Grimm, C., Herz, A., Tulunay, F.C. & Loh, H.H. (1982a). Life Sci. 31, 1883-1886.
Höllt, V., Tulunay, F.C., Woo, S.K., Loh, H.H. & Herz, A. (1982b). Europ. J. Pharmacol. 85, 355-356.
Höllt, V. (1983). Trends in Neurosci. 6, 24-26.
Huidobro-Toro, J.P., Yoshimura, K., Lee, N.M., Loh, H.H. & Way, E.L. (1981). Europ. J. Pharmacol. 72, 265-266.
Kakidani, H., Furutani, Y., Takahashi, H., Noda, M., Morimoto, Y., Hirose, T., Asai, M., Inayama, S., Nakanishi, S. & Numa, S. (1982). Nature 298, 245-249.
Kangawa, K., Minamino, N. Chino, N., Sakakibara, S. & Matsuo, H. (1981). Biochem. biophys. Res. Commun. 99, 871-878.
Khachaturian, H., Lewis, M.E., Höllt, V. & Watson, S.J. (1983). J. Neurosci. 3, 844-855.

Kilpatrick, D.L., Taniguchi, T., Jones, B.M., Stern, A.S., Shively, J.E., Hullihan, J., Kimura, S., Stein, S. & Udenfriend, S. (1981). Proc. Natl. Acad. Sci. USA 78, 3265-3268.
Kilpatrick, D.L., Wahlström, A., Lahm, H.W., Blacher, R. & Udenfriend, S. (1982). Proc. Natl. Acad. Sci. USA 79, 6480-6483.
McKnight, A.T., Corbett, A.D., Paterson, S.J., Magnan, J. & Kosterlitz, H.W. (1982). Life Sci. 31, 1725-1728.
McKnight, A.T., Corbett, A.D. & Kosterlitz, H.W. (1983). Europ. J. Pharmacol. 86, 393-402.
Kosterlitz, H.W., Lord, J.A.H., Paterson, S.J. & Waterfield, A.A. (1980). Br. J. Pharmac. 68, 333-342.
Kosterlitz, H.W., Paterson, S.J. & Robson, L.E. (1981). Br. J. Pharmac. 73, 939-949.
Lord, J.A., Waterfield, A.A., Hughes, J. & Kosterlitz, H.W. (1977). Nature 267, 495-499.
Martin, W.R., Eades, C.G., Thompson, J.A., Huppler, R.E. & Gilbert, P.E. (1976). J. Pharmacol. exp. Ther. 197, 517-532.
Minamino, N., Kangawa, K., Fukuda, A. & Matsuo, H. (1980). Biochem. biophys. Res. Commun. 95, 1475-1481.
Minamino, N., Kangawa, K., Chino, N., Sakakibara, S. & Matsuo, H. (1981). Biochem. biophys. Res. Commun. 99, 864-870.
Mizuno, K., Minamino, N., Kangawa, K. & Matsuo, H. (1980a). Biochem. biophys. Res. Commun. 95, 1482-1488.
Mizuno, K., Minamino, N., Kangawa, K. & Matsuo, H. (1980b). Biochem. biophys. Res. Commun. 97, 1283-1290.
Nakanishi, S., Inoue, A., Kita, T., Nakamura, M., Chang, A.C.Y., Cohen, S.N. & Numa, S. (1979). Nature 278, 423-427.
Noda, M., Furutani, Y., Takahashi, H., Toyosato, M., Hirose, T., Inayama, S., Nakanishi, S. & Numa, S. (1982). Nature 295, 202-206.
Oka, T., Negishi, K., Suda, M., Matsumiya, T., Inazu, T. & Ueki, M. (1980). Europ. J. Pharmacol. 73, 235-236.
Oka, T., Negishi, K., Kajiwara, M., Watanabe, Y., Ishizuka, Y. & Matsumiya, T. (1982). Europ. J. Pharmacol. 79, 301-305.

Rezvani, A., Höllt, V. & Way, E.L. (1983). Life Sci., in press.
Schulz, R., Wüster, M. & Herz, A. (1982). Peptides 3, 973-976.
Wüster, M., Schulz, R. & Herz, A. (1980). Europ. J. Pharmacol. 62, 235-236.
Wüster, M., Schulz, R. & Herz, A. (1981a). Biochem. Pharmacol. 30, 1883-1887.
Wüster, M., Rubini, P. & Schulz, R. (1981b). Life Sci. 29, 1219-1227.

PART II

Novel Neuropeptides

STRUCTURE AND FUNCTION OF A NEW NEUROPEPTIDE, THE HEAD ACTIVATOR, CONTROLLING HEAD FORMATION IN HYDRA

H. Chica Schaller and
Heinz Bodenmüller *

Department of Biophysics
Max-Planck-Institute for Medical Research
Heidelberg, FRG

From the freshwater coelenterate hydra we recently isolated and sequenced a peptide which acts as a head inducing morphogen. This peptide, called head activator, is responsible in hydra for head-specific growth and differentiation processes. It affects growth by triggering cells to divide and cellular differentiation by influencing the determination of cells. A peptide with the same amino acid sequence as the head activator was also isolated from mammalian brain and intestine. Immunological methods indicate that the head activator is especially abundant in frog skin and in tumors of APUD origin. This is the first example that a neuropeptide conserved its structure and may be its trophic function all through evolution from coelenterates to humans.

I. INTRODUCTION

Within the animal kingdom cells specialised for nervous function at first appear among coelenterates. We have chosen to work with hydra, because its nervous system and its general organisation seemed to be simple. The animal consists of only two cell layers, ectoderm and entoderm, among which luxury cells such as nerve cells and others are interspersed. The nervous system is organised as a continuous net with a higher density in the head and in the foot region.

* Address for correspondence: Department of Biophysics, Max-Planck-Institute for Medical Research, Jahnstr. 29, 6900 Heidelberg, FRG

ISBN 0-12-417320-9

It serves two functions: (1) It is necessary for the coordination of fast processes like body movements, and (2) it controls slow processes like cellular growth and differentiation. Our aim is to understand how this nervous system functions, in particular, to analyse the substances involved in intercellular communication.

So far not much is known how and with which transmitter substances the fast interneuronal communication system in hydra works. Accumulating evidence suggests that conventional transmitters such as acetylcholine (1) or catecholamines are not present in nerve cells of hydra. It seems more likely that peptides may play that role since more and more of them are found coexisting or separately in nerve cells of hydra (2). We know more about the second function of the nervous system and about the substances which control cellular growth and differentiation and thus pattern formation in hydra. To understand pattern formation we want to know (1) why a certain structure is induced at a definite position and time during development, and (2) why it subsequently inhibits the formation of an identical such structure nearby. Hydra has two centers of organisation, head and foot, from both of which gradients of induction and inhibition extend towards the opposite ends. This has been deduced mostly from transplantation, but also from other biological experiments. One basic experiment is the following: If a hydra is cut into two pieces of equal length each piece will regenerate the missing structure such that two perfect hydra result. The cut surface is identical in both cases. However, from the same piece of tissue in the upper half a foot is regenerated, in the lower half a head. Transplantation experiments have shown that the induction of the respective structures, head or foot, is a relatively fast process, that within a few hours after cutting (3 - 6 h) the regenerating gastric tissue has changed its program with the result that is behaves more like a foot or a head tissue (3). Induction is a very local process and restricted to the tissue close to the original cut. At the cellular level this means that only cells close to the original cut surface are changed in their growth and differentiation characteristics, i.e. become head-specific of foot-specific cells, respectively. Inhibition on the other hand is long range: an existing head usually inhibits that another head is formed along the whole body column. Only if a certain size is reached, if the distance to the existing head is large enough, a secondary head induction is allowed. The potency to induce or inhibit a head is highest in or close to the head region decreasing in a graded fashion to the foot. Similarly, the potency of foot induction or foot inhibition decreases from foot to head.

Our contribution to a better understanding of the phenomena of induction and inhibition in hydra is that we could show that the gradients in biological properties are caused by a graded distribution along the body axis of the animal of four morphogenetic substances that induce or inhibit head or foot formation (for a review see 4). Recently we succeeded in isolating and sequencing one of those, namely a peptide which acts as a head-specific morphogen in hydra - the head activator (5). We also isolated this peptide from rat intestine and from human hypothalamus and found that the structure is identical (6). In the following some of the chemical and biological properties of this molecule are discussed and an idea is presented how we think it functions.

II. ISOLATION AND SEQUENCE ANALYSIS OF THE HEAD ACTIVATOR

The sources, from which the head activator was isolated and sequenced, are listed in Table 1.

TABLE 1: Head activator yield from different animal sources

Animal source	Starting material	Yield of head activator
Coelenterates:		
Hydra attenuate	3 kg	500 pmol (0.5 μg)
Anthopleura elegantissima	200 kg	20 nmol (20 μg)
Mammals:		
Rat intestine	3000 pieces	20 nmol (20 μg)
Human hypothalamus	2 pieces	2 nmol (2 μg)
Bovine hypothalamus	10 kg	10 nmol (10 μg)

To obtain workable quantities of the peptide we had to start with relatively large amounts of tissue. In general, we could enrich the head activator 10^8-10^9-fold with a 10-20% yield. Purification steps included organic solvent extraction, molecular sieve and ion exchange chromatography, and HPLC methods (5). The head activator was located in the various purification steps and column fractions by assaying the biological activity as stimulation of bud outgrowth in hydra (7). In later steps, when the head-activator peptide was free from other protein or peptide contaminants, we usually located it first by hydrolysing a small sample of all the fractions and looked for the characteristic head-activator pattern of the dansylated amino acids on thin layer plates. Ten picomoles were sufficient for that purpose. In a second step we then determined the biological activity. As final criterium for purity and idendity of the peptide we used reverse-phase HPLC. In combination with a high sensitivity UV detector HPLC makes possible the detection of small amounts (100 pm) of peptides. If at least two different solvent systems are used and the peptide elutes as a sharp peak one can be reasonably sure that it is pure. We found that peptides isolated from the different animal sources were identical in their chromatographic behavior. This meanwhile includes head activator isolated from frog skin, and from various mammalian tumors or tumor cell lines derived from cells of neural crest origin.

The amounts of head activator available for a sequence analysis were relatively minute and required use and improvement of micromethods. To determine the amino acid composition of the head activator the purified peptide was hydrolysed, dansylated and the dansylated amino acids were separated for a qualitative analysis on micro thin-layer plates and for quantitation on HPLC (5). We found that the head activator contained the following amino acids: Gly (2), Glu (1), Ile (1), Leu (1), Lys (1), Phe (1), Pro (2), Ser (1), Val (1). The sequence analysis was complicated by the fact that the head activator had no free amino terminus, that its amino end was blocked by pyroglutamic acid. Therefore, most of the sequence was obtained by enzymatic degradation methods and in part by a chemical degradation of the biologically inactive Glu-peptide which contained glutamic acid instead of pyroglutamic acid as amino terminal amino acid (5). This Glu-peptide was produced probably as an artifact of the HPLC procedure. The head activator was fragmented by an endopeptidase isolated from *Astacus fluviatilis* (AEP) (7). Incubations of the pGlu-peptide with AEP yielded two fragments and Gly as free amino acid (5). In Table 2 the different sequencing steps and their results are outlined.

TABLE 2: Sequence analysis of fragments of the head activator.

Method	Amino acid sequence
Edman degradation	Glu-Pro-Pro-Gly-Gly-
Carboxypeptidase Y	-Lys-Val-Ile-Leu-Phe
AEP	-Gly-
AEP and end group	-Gly-Ser-
AEP and Edman	Glu-Pro-Pro-Gly-
AEP and aminopeptidase M	-Gly-Ser-Lys-Val-Ile-Leu-Phe

The head activator was found to have the sequence

pGlu-Pro-Pro-Gly-Gly-Ser-Lys-Val-Ile-Leu-Phe.

No difference in sequence was found for the head activator from the sea anemone, *Anthopleura elegantissima*, from *Hydra attenuata*, from rat intestine and from bovine and human hypothalamus. This sequence of the head activator was confirmed by synthesis (8). We could show that there is no difference between the purified native and the synthetic peptides in chemical, enzymatic and biological properties. We and others also produced a series of peptides with minor modifications. So far none of those was as biologically active as the native molecule. Together with the evolutionary conservation of the head activator from hydra to mammals we interprete this to mean that the native molecule has very strict conformational requirements.

The chemistry of the head activator predicts or explains the following outstanding properties: (1) It is very lipophilic leading to a relatively low solubility in aqueous solvents. (2) In common with many other biologically active peptides it has a very high surface adhesivity resulting in absorption in vitro to glass, plastic etc and in vivo probably to membranes and to hydrophobic parts of proteins rendering it relatively immobile or preventing diffusion to distant areas. (3) The head activator occurs in two distinguishable conformations, one with an apparent molecular weight of about 1500 daltons and one of about 700 daltons. Preliminary evidence supports the idea that the l.m.w. conformation is a monomeric peptide, whereas the larger one is dimeric. The freshly isolated, relatively impure, and dilute native peptide has the monomeric conformation. The synthetic molecule and the purified head activator in physiological media converts to the more stable dimeric conformation. The two conformations are interconvertible. Since only the monomer is biologically active we think that this conversion may be of physiological relevance. (4) The head activator is a poor immunogen. This may be due to the fact that it is an inherent or endogenous molecule and for that it has very few antigenic determinants, especially if one or both of the two functional groups is used for coupling the peptide as hapten to a higher molecular weight carrier for immunisation.

III. EFFECTS OF THE HEAD ACTIVATOR ON MORPHOGENESIS, CELLULAR GROWTH AND DIFFERENTIATION

The head activator and its counterpart, the head inhibitor, are present in hydra in an inactive structure-bound form. The head activator is stored in neurosecretory granules (9). This has the advantage that the release can be regulated. The release of the head factors is controlled by the head inhibitor which acts as a release-inhibiting factor for its own release and for that of the head activator as well. The head activator has no effect on its own release nor on that of the head inhibitor. Thus release of both factors is triggered by the absence of the head inhibitor. The head inhibitor is a small molecule of around or less than 500 daltons, it contains no peptide bonds, and it is very hydrophilic. Based on those chemical properties and as demanded by biological evidence the head inhibitor is able to diffuse, once released into the intercellular space, over long distances. The head activator due to its hydrophobic character will adhere to membranes of neighboring cells or to the extracellular matrix and will therefore be restricted locally. The distribution of sources for both substances within hydra is relatively similar, both having their maximal concentration in the head area (hypostome). Thus although both substances may be released from the same location, at the release site there will be a local predominance of head activator over head inhibitor and at a greater distance predominance of head inhibitor over head activator. At the cellular level head activator and head inhibitor act antagonistically: The head activator stimulates cells to divide and it is necessary for head-specific differentiation, the head inhibitor blocks such processes (10-12). Thus an interstitial stem cell will become a head-specific nerve cell if the concentration of the head activator is higher than that of the head inhibitor. The stem cell will remain stem cell or enter another differentiation pathway (e.g. to nematocytes), if head inhibitor prevails. This means that close to a common release site stem cells differentiate to nerve cells whereas further away they enter another differentiation pathway. The new nerve cells will then produce head activator and/or head inhibitor thus reinforcing or amplifying the head-specific influence. Such a feedback mechanism may ensure that wherever head activator predominates over head inhibitor, head-like structures will be induced and maintained.

IV. ACTION OF THE HEAD ACTIVATOR IN MAMMALS

The sequence of the head activator differs from that of other biologically active peptides. The only structural homology we observed is with bradykinins, in the sequence X-Pro-Pro-Gly-X-Ser-. Bradykinins are known as smooth muscle stimulants. The head activator at a concentration of 10^{-5} M did stimulate uterus contraction in rats (Ganten, unpublished). This may imply a very weak structural and functional homology. From the localisation of the head activator in the brain and intestinal tract, from its appearance early in mammalian development and in analogy to its action in hydra, the following speculation on its action may be allowed. Like other neuropeptides it may act as a transmitter-like messenger in the brain. The high concentration in the intestine may hint at a regulatory function in the digestive system. In embryonal life and may be also in the adult the head activator may act on some neuronal stem cell either as growth factor or as differentiation signal. Preliminary evidence suggests that some of these ideas may be correct. Much remains to be done to substantiate the role of the head activator in such functions in mammals.

ACKNOWLEDGEMENT:

H.C. Schaller is a recipient of a Heisenberg fellowship, we are supported by the Deutsche Forschungsgemeinschaft (Scha 253/8), and by the Fond der deutschen chemischen Industrie.

REFERENCES

1. Erzen, J. and Brzin, M. (1978) Comp. Biochem. Physiol 59c, 39-43.
2. Grimmelikhuijzen, C.J.P., Dierickx, K. and Boer, G.J. (1982) Neuroscience 7, 3191-3199.
3. Berking, S. (1979), Wilhelm Roux's Archives 186 189-210.
4. Grimmelikhuijzen, C.J.P. and Schaller, H.C. (1979) Trends in Biochemical Sciences 4, 265-267.
5. Schaller, H.C. and Bodenmüller, H. (1981) PNAS 78, 7000-7004.
6. Bodenmüller, H. and Schaller, H.C. (1981) Nature 293, 579-580.
7. Zwilling, R., Dörsam, H., Torff, H.-J., and Rödl, J. (1981), FEBS Lett. 127, 75-78.
8. Birr, C., Zachmann, B., Bodenmüller, H., and Schaller, H.C. (1981), FEBS Lett. 131, 317-321.
9. Schaller, H.C. (1979), TINS, , 120-122.
10. Berking, S. (1979), J. Cell Sci. 40, 193-205.
11. Schaller, H.C. (1976), Cell Differ. 5, 1-11.
12. Schaller, H.C. (1976), Cell Differ, 5, 13-20.

ISOLATION AND NH_2-TERMINAL SEQUENCE OF A HIGHLY CONSERVED PITUITARY AND HYPOTHALAMIC PROTEIN BELONGING TO A NEW SUPERFAMILY

Michel Chrétien[1]
Kuo-Liang Hsi[2]
Nabil G. Seidah

Protein and Pituitary Hormone Laboratory
Clinical Research Institute of Montreal
Montreal, Canada

During the course of purification of the N-terminal fragment of pro-opiomelanocortin we isolated and purified a 21,000 daltons (pI 4.9) novel protein from porcine anterior pituitaries and whole human pituitary glands whose amino acid composition revealed a high homology between these two species. The NH_2-terminal sequence of the first 77 residues of the human and porcine homologues confirmed the homology since one conservative substitution was found at residue 12, namely an Ala for a Thr. A computer data bank search using a mutation data matrix and comparison with 338,327 segments of proteins revealed that this substance should be classified as belonging to a new protein superfamily. An antibody produced against a synthetic fragment of this molecule was raised and immunocytochemical staining revealed the presence of material in the anterior and posterior lobe of the pituitary and in the supraoptic nucleus of the hypothalamus. It was absent in the intermediate lobe. This novel pituitary and hypothalamus substance has partial sequence homology with duck pro-insulin, Rous sarcoma virus protein and secretin.

[1] *This paper is dedicated to Margaret Dayhoff.*
Work supported by the Medical Research Council of Canada, the National Institutes of Health (MS016315-03) and the National Cancer Institute of Canada.
[2] *Present Address: Shanghai Institute of Materia Medica, China*

ISBN 0-12-417320-0

I. PITUITARY PROTEINS AND HORMONES

The anterior and the posterior pituitary lobes (1-3) are sources of many hormones, proteins and polypeptides (4-7). The secretory material found in the posterior pituitary are, however, synthesized in specific nuclei of the hypothalamus and reach the posterior pituitary via axonal transport and are stored in nerve terminals (4).

The possibility that the pituitary might contain as yet undiscovered regulatory substances is intriguing. Recent studies have identified many polypeptides in the pituitary and suggested that they may play a role therein. Good examples are angiotensin II (8), substance P (9), epidermal growth factor (10), and α-2u-globulin (11-12).

We have recently discovered a protein which bears some sequence homology to duck pro-insulin, Rous sarcoma virus protein and pig secretin (13). In this report, we describe the amino acid sequence of the porcine protein to be 81 residues and we have sequenced the first 77 residues of its human homologue. Results show that this molecule is highly conserved (98.7%) between those two species within the determined sequence. An exhaustive computer data bank search revealed that this is a novel protein belonging to a new superfamily. Furthermore, immunocytochemical studies showed the presence of this substance in the pars nervosa and pars distalis of the pituitary, and its absence from the pars intermedia. In addition, it was localized in nerve fiber varicosities within the hypothalamus of human, rat and mouse brain.

II. MATERIALS AND METHODS

A. *Isolation and Purification of Human and Porcine Homologues*

Following HCl/acetone extraction of the glands (14), the acid-acetone powder thus obtained was dissolved in water and the solution brought to 6% (w/v) NaCl and the pH adjusted to 3.0. The precipitate was desalted and lyophilized. This crude material was subjected to preparative reverse-phase high performance liquid chromatography (HPLC) on a μ-Bondapak C-18 column (0.70 x 30 cm) (Waters) using a linear gradient of 2-propanol in 0.13% (v/v) heptafluorobutyric acid (HFBA). The material from this first purification step represented about 0.1% of the starting gross extract (Fig. 1A and 1D).

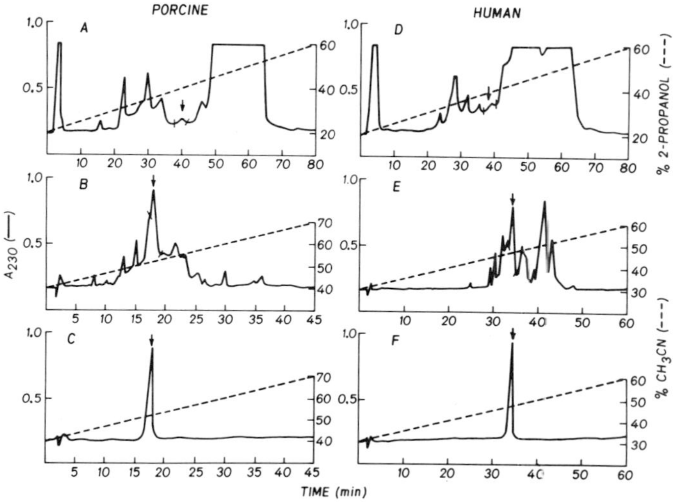

FIGURE 1. Reverse phase HPLC purification of the whole human and anterior porcine pituitary extracts. Chromatograms A and D depict the purification of the initial extract on a semi-preparative μ-Bondapak C-18 column (0.7 x 30 cm) equilibrated with 20% 2-propanol (containing 0.13% v/v HFBA). The linear gradient used was 20% to 60% 2-propanol in 80 min at a flow rate of 2 ml/min. Material under the arrow was collected automatically for further purification, as shown in panels B, C, E and F. In panels B and E, elution was made on an analytical μ-Bondapak C-18 column (0.39 x 30 cm) equilibrated with either 40% or 30% CH_3CN (containing 0.1% v/v TFA) respectively. The linear gradient consisted of either 40% to 70% CH_3CN (panel B) or 30% to 60% CH_3CN (panel E) both at a flow rate of 1 ml/min. The material under each peak was repurified on a similar analytical column under the same conditions (panels C and F). The dashed lines represent the linear gradient used in each case. In all cases the elution was made at room temperature and was monitored by the absorbance at 230 nm.

This first fraction was then repurified on an analytical μ-Bondapak C-18 column (0.39 x 30 cm) using a linear gradient of acetonitrile of 0.1% (v/v) trifluoroacetic acid (TFA) (Fig. 1B and 1E). The material of interest represented 10% by weight of the first fractionation. The final repurification was done on a similar column using the same gradient of acetonitrile in 0.1% TFA (see Fig. 1C and 1F).

B. *Molecular Weight and Isoelectric Point Estimation*

The molecular weight of the purified human and porcine homologues was estimated using slab sodium dodecyl sulfate-polyacrylamide gel electrophoresis (SDS/PAGE) at three different acrylamide concentrations (10-15%, 10-20% and 15-25%) (13). In all cases both human and porcine proteins migrated with an apparent molecular weight of 21,000. Two-dimensional SDS-PAGE/isoelectric focusing was performed (15) and both proteins exhibited a single band migrating with an apparent isoelectric point of 4.9.

C. *Amino Acid and Microsequence Analyses*

Triplicate amino acid analysis of reduced and carboxy-methylated proteins hydrolysed for 24, 48 and 72 hrs in 5.7N HCl at 108°C under vacuum were done on an updated Beckman 120C amino acid analyzer (Table 1). The cyanogen bromide fragments obtained from the reduced and carboxy-methylated porcine material were analyzed following duplicate 24 hrs hydrolysis. Starting from 10 nmoles of the reduced and carboxymethylated human protein, automated liquid phase Edman degradation was performed using a 0.3M Quadrol program and 3 mg Polybrene (Aldrich) (16), on a Beckman 890C sequencer equipped with a Sequemat P-6 autoconverter and a model SC-510 controler. The sequence of the cyanogen bromide fragments of the porcine material was also performed in a similar fashion.

D. *Cyanogen Bromide Cleavage of the Porcine Homologue*

CNBr cleavage of the reduced and carboxymethylated porcine protein (Fig. 1C) was made in 70% formic acid with a molar ratio of CNBr to methionine of 50 to 1. Reaction was allowed to proceed at room temperature for 24 hrs in the dark (16). After lyophilization, the cleavage products were then purified on a μ-Bondapak C-18 column (0.39 x 30 cm) eluted with a linear gradient of 2-propanol in 0.13% (v/v) HFBA (Fig. 2).

Table 1. Composition of Native Material and Fragments

Amino Acids	Human	Porcine	Porcine CNBr fragments		
			CNBrI	CNBrII	CNBrIII
Asp	20.71 (21)	19.92 (20)	2.09 (2)	0 (0)	7.87 (8)
Thr	7.09 (7)	7.05 (7)	1.88 (2)	0 (0)	1.83 (2)
Ser	10.30 (10)	7.16 (7)	1.70 (2)	0 (0)	2.71 (3)
Glu	25.99 (26)	26.40 (26)	2.40 (2)	4.10 (4)	9.86 (10)
Pro	16.67 (17)	16.84 (17)	2.06 (2)	2.11 (2)	6.81 (7)
Gly	17.26 (17)	16.64 (17)	1.29 (1)	1.12 (1)	7.79 (8)
Ala	11.67 (12)	11.72 (12)	0 (0)	2.93 (3)	4.08 (4)
Cys[1]	3.78 (4)	4.49 (4)	0 (0)	0 (0)	1.47 (1)
Val	7.85 (8)	7.08 (7)	1.79 (2)	0.98 (1)	2.76 (3)
Met	2.98 (3)	2.70 (3)	1[2] (1)	1[2] (1)	1[2] (1)
Ile	6.44 (6)	7.08 (7)	0.90 (1)	1.00 (1)	3.78 (4)
Leu	16.43 (17)	16.00 (16)	1.91 (2)	1.09 (1)	5.70 (6)
Tyr	5.60 (6)	5.77 (6)	0.95 (1)	1.11 (1)	1.83 (2)
Phe	5.95 (6)	5.75 (6)	0 (0)	0 (0)	2.15 (2)
Lys	9.02 (9)	8.78 (9)	0 (0)	0 (0)	1.66 (2)
His	7.15 (7)	6.76 (7)	0.97 (1)	0.99 (1)	3.06 (3)
Arg	7.83 (8)	8.35 (8)	2.65 (3)	1.91 (2)	2.03 (2)
Trp[3]	n.d.	n.d.	n.d.	n.d.	n.d.
GlcN[4]	0 (0)	0 (0)	0 (0)	0 (0)	0 (0)
GalN[5]	0 (0)	0 (0)	0 (0)	0 (0)	0 (0)
Presumed Total	183[6]	179[6]	22	18	68[6]

[1] Determined as carboxymethyl cysteine
[2] Determined as sum of homoserine and homoserine lactone
[3] n.d. means not determined
[4] GlcN means glucosamine
[5] GalN means galactosamine
[6] The exact composition will need to be confirmed by the complete sequence

Amino acid composition of human and porcine reduced and carboxymethylated proteins, together with that of three cyanogen bromide fragments CNBrI, II, and III. The values given for the whole protein are obtained following triplicate analysis of 24, 48 and 72 hrs hydrolysates. The cyanogen bromide amino acid analysis was done only on 24 hrs hydrolysates. It can be seen that this protein is not glycosylated in view of the absence of glucosamine and galactosamine in all the amino acid analysis performed on native and cyanogen bromide fragments. The amino acid analysis of CNBrIII represents the minimum composition that could be deduced for this peptide based on the determined sequence of the segment 41-81 of the molecule.

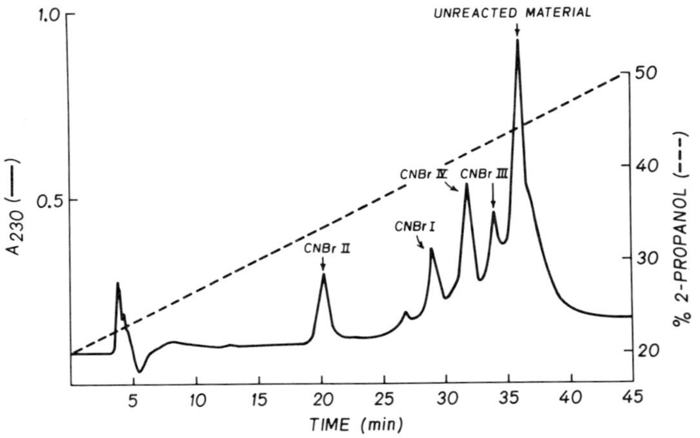

FIGURE 2. HPLC separation of the cyanogen bromide fragments of the porcine analogue. Here an analytical μ-Bondapak column eluted with a linear gradient (dashed line) of 2-propanol/0.13% HFBA was used at a flow rate of 1.5 ml/min. The four cyanogen bromide fragments obtained are denoted CNBrI to CNBrIV. A large amount of the material resisted CNBr cleavage indicating oxidation of the methionines in the molecule.

E. Antibody Production Against a Segment of the Protein

For immunization, a synthetic peptide containing amino acid 23-29 was coupled to thyroglobulin, at a mole ratio of 10 to 1 using a water- soluble carbodiimide (EDAC, Biorad). Following three boosts, the titer in one of the rabbits was found to be 1/10,000. This provided the antibody source for the immunocytochemical localization of the protein. This antibody was found to cross react (20%) with the native porcine material, but does not recognize any pro-opiomelanocortin (3,16) segment including the N-terminal (16), ACTH, α-MSH, β-LPH or β-endorphin. Furthermore, no cross reaction was found with neurophysins, vasopressin, oxytocin, and growth hormone.

III. RESULTS

A. *Isolation and Purification of the Human and Porcine Homologues*

Fig. 1 shows the purification scheme of both human and porcine material. As can be seen, both proteins elute at similar positions on all HPLC systems used. Furthermore, they elute at a relatively high concentration of either 2-propanol/HFBA (around 40%) or acetonitrile/TFA (around 44%). The amount of material that can be obtained is quite small. For example, out of 10 g of porcine anterior pituitary extract (about 1000 pituitaries) we could only purify 5 mg of this material, giving a yield of about 0.05% overall. Also, about 2 mg was purified out of 2 g of whole human pituitary extract, giving a 0.1% overall yield. The difficulty of separating it from growth hormone on either molecular sieving or carboxymethyl cellulose chromatography, led the way to the use of reverse phase HPLC as a good procedure to achieve such a purification. In the systems used, growth hormone elutes much later on the column (around 50-55% 2-propanol). The HPLC collection system was, therefore, automated permitting the processing of 10 g of pituitary extract within 10 days of continuous runs. The fraction collected (see Fig. 1A and 1D) was then rapidly repurified in two acetonitrile/TFA gradient systems (see Fig. 1B, C, E, F).

The homogeneity of the material was verified by SDS/PAGE and isoelectric focusing whereby only one band was obtained for both homologues (13), exhibiting an apparent molecular weight of 21,000 and an isoelectric point of pH 4.9.

These proteins were then reduced and carboxymethylated. Their amino acid compositions are given in Table 1. It can be seen that the amino acid composition of the human and porcine homologues are quite similar, indicating a possible high degree of homology between the two species. Furthermore, it is clear that these two proteins are not glycosylated. The presence of 4 cysteines would predict two disulfide bridges. With respect to the total number of amino acids present in the molecule, the values given (183 and 179 for the human and porcine homologues respectively) are based on the apparent molecular weight of the molecule, and as shown later, on the results of the sequence of the first 81 residues. However, only the total sequence will determine their exact length.

B. *Amino-terminal Sequence of the Human and Porcine Homologues*

Starting from 10 nmoles of the human carboxymethylated protein, automatic microsequencing allowed the unambiguous determination of the primary structure of the first 77 residues. The overall repetitive yield of this sequence was found to be about 94.5%. In preliminary sequence runs, it was found that the yield dropped sharply following certain Pro residues, thus it was decided to automatically perform double cleavages at every proline residue during its sequential degradation on the sequenator. This definitely improved the overall yield and the carry-overs between successive cycles.

The same strategy was also employed for the porcine homologue and the first 50 residues were unambiguously determined. In this case, the overall repetitive yield, starting from 25 nmoles of the reduced and carboxymethylated protein, was found to be around 97%. In order to extend the sequence of the porcine homologue, the presence of 3 Met in the molecule (Table 1), suggested the use of CNBr as the cleavage procedure. In Fig. 2, the HPLC separation of the CNBr fragments generated is depicted. It was found that 4 fragments could be generated, as expected from the presence of 3 methionine residues in the molecule (Table 1). The amino acid composition of the CNBrI, CNBrII and CNBrIII fragments is given in Table 1. The sequence of CNBrI and CNBrII confirmed their identity as representing the amino terminal segments 1-22 and 23-40 of the molecule, in agreement with the sequence of the native polypeptide. Furthermore, the amino acid composition of these fragments fits exactly the determined sequence. With respect to the CNBrIII fragment, its composition predicts a probable length of 68 residues (Table 1). This is a minimum since its sequence (Fig. 3) shows 4 isoleucines within the first 41 residues of this fragment, in agreement with its composition. The exact length of this molecule will have to be ascertained by complete sequence. As far as the CNBrIV fragment is concerned, its limited availability and its heterogeneity on HPLC made it impossible to fully analyze it.

The summary of these sequence results is given in Fig. 3. The NH_2-terminal 50 residues sequence of the porcine analogue, in addition to the sequence of CNBrI, CNBrII and the first 41 residues of CNBrIII, provided the necessary sequence overlaps and confirmation to allow the proposal of the sequence of the first 81 residues of the porcine protein. The comparison of this sequence with that of the human homologue is also shown

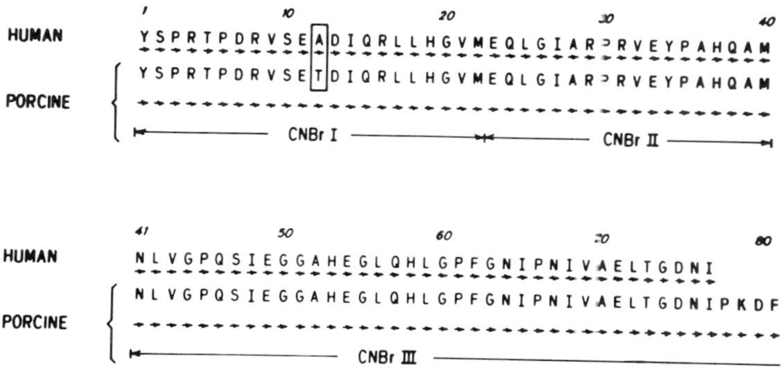

FIGURE 3. The deduced amino acid sequence of the first 77 and 81 residues of the human and porcine homologues respectively. The only variation found between the two species at residue 12 is boxed in. The extension of the porcine sequence to 81 residues was made possible by the CNBrIII fragment starting at residue 41 of the molecule. This provided the overlap at residue 50-51 (13) and allowed the extension of the sequence to residue 81.

in Fig. 3. Surprisingly, only one variation exists between the two species within the first 77 residues, namely an Ala instead of a Thr at residue 12 in the human and porcine homologues respectively. This substitution is very conservative since it involves only one nucleotide change in all codons coding for either Ala or Thr (namely a guanine for an adenine in the first nucleotide of the triplets). This high degree of sequence conservation (98.7% within the first 77 residues), reflected also in the overall amino acid composition of both proteins, would indicate a high degree of pressure during evolution to resist change, and could predict an important biological function.

C. *Computer Data Bank Search for Possible Sequence Similarities with Other Proteins or Segments of Proteins*

We have already shown in a previous report (13) that, based on the 50 amino terminal sequence residues of the porcine homologue, an exhaustive computer data bank search (performed by the late Dr. Margaret Dayhoff of the National Biomedical Research Foundation, Georgetown University, Washington D.C.) revealed that less than 30% homology could be found with any known protein or segment of protein. However, such mutation data matrix search did reveal suggestive homology with duck pro-insulin (30%), Rous Sarcoma virus transforming protein TVFV60 (24%), and pig secretin (26%) (13). With the extended sequence reported in this paper, such a search was repeated. Here also, Dr. M. Dayhoff showed that out of 338,327 segments of proteins searched representing a total of 2030 protein sequences, no significant match could be found, except for the proteins previously reported (13). However, this match is now inferior to 25% and suggests a membership to a new superfamily. Furthermore, back translation of all the known coding regions from the nucleic acid sequence data base did not show a significant match out of more than two million codons searched, again reinforcing the novelty of such a sequence. In addition, from the amino acid composition of the human and porcine homologues (Table 1), a match search was initiated which also revealed no significant homology to the composition of any known protein or segment thereof. This is quite significant since Dr. Dayhoff has shown that if more than 50% homology exists between two or more proteins the amino acid composition search would surely pick-up such a similarity and could suffice to identify the protein. In conclusion, this computer data bank search revealed that this highly conserved protein is indeed novel, and should be classified as belonging to an entirely new superfamily.

D. *Tissue Distribution*

Using an antiserum produced in rabbits against a synthetic fragment representing the sequence 23-39 of either the human or porcine homologues, an immunocytochemical study was performed in order to localize this protein in pituitary and hypothalamus of human, rat and mouse. The specific staining was found in the anterior and posterior lobes of the pituitary and also in the hypothalamic neurons of the rat, identified as belonging to the supraoptic nucleus. However, no detectable immunoreactivity could be seen in the intermediate lobe of the rat and mouse pituitary. Discrete varicose nerve fibers of the human hypothalamus were also immunocytochemically stained (not shown). Specificity of the immunocytochemical staining was demonstrated by the preabsorption of this antiserum with variable amounts of either the synthetic fragment or the native porcine homologue resulting in near complete absence of staining.

IV. DISCUSSION

A novel highly conserved protein has been isolated in minute amounts from the anterior pituitary of pig and whole human pituitary extracts. This protein belonging to an entirely new superfamily turned out by immunocytochemical staining to be found in the anterior and posterior lobes of the hypophysis, but absent in the pars intermedia (data not shown). Its presence in the pars nervosa of the pituitary led to the hypothesis that it could be synthesized in the hypothalamus and transported through axons into the posterior lobe of the pituitary. Such a case is well known for the vasopressin/oxytocin family of peptides and their associated neurophysins (4,6). Indeed, specific immunocytochemical staining was found in discrete regions of the hypothalamus including nerve fibers in the median eminence and nerve cell bodies belonging to the supraoptic nucleus. Furthermore, if such is the case one would expect to find this substance or fragment thereof in posterior pituitary secretory granules. This was confirmed as found by radioimmunoassay of porcine neurointermediate pituitary secretory granule preparations (unpublished results). With respect to the distribution of this protein in the anterior and posterior lobes of the pituitary and in the supraoptic nucleus of the hypothalamus, a similar type of distribution was reported for the opioid peptide dynorphin (7,20). However, for dynorphin the amounts found in the posterior pituitary were about 22 times higher than those found in the anterior pituitary (21). Furthermore, the molecular weight of dynorphin-like material in the anterior pituitary was about 6,000 as compared to 2,000 in the posterior pituitary (20). It cannot be excluded at present that the sizes of the isolated protein are different in the two lobes due to different post-translational processing. Different processing of a pro-hormone in particular tissues have already been demonstrated for pro-opiomelanocortin (POMC) which is enzymatically cleaved into higher molecular weight forms, β-LPH and ACTH within the adenohypophysis, while these two peptides are further processed into smaller forms, i.e., β-endorphin and α-MSH within the pars intermedia (3).

The conservation of the sequence of segment 23-39 of the molecule between species is indicated by the possibility of using an antiserum to the synthetic human and porcine 23-39 sequence and detecting an immunostaining reaction in the pituitary of rat and mouse, as well as human.

The fact that this protein was localized in the anterior pituitary cells, largely in the cytoplasm, could indicate its synthesis therein or its uptake from the circulation, as found for a number of neuropeptides (22) and peripheral hormones. However, the porcine material was purified from the anterior lobe in quantities largely exceeding those of known peripheral or neurohormones; this would support the view that this protein is synthesized, at least in part, in the anterior lobe of the pituitary. The possibility that the posterior pituitary protein localized by immunocytochemistry is of hypothalamic origin is evidenced by its localization in discrete nerve fibers and neuronal cell bodies in the hypothalamus, and the known absence of neuronal cell bodies in the posterior lobe of the pituitary. Current studies are being done to further identify other possible specific loci within the brain.

With regards to its biological role, it is yet too early to define it. However, the homology with pro-insulin and Rous Sarcoma virus transforming protein suggested that it could function as a growth factor. Such a possibility is presently under investigation, together with a more detailed distribution of this peptide in various tissues and organelles.

ACKNOWLEDGMENTS

The authors thank G. De Serres, J. Rochemont and J. Hamelin for their collaboration. They express their gratitude to Drs. H. Iguchi and J.S.D. Chan for their help in the development of the antiserum and the set up of the radioimmunoassay; Dr. D. Pélaprat for the preparation of the neurosecretory granules from porcine neurointermediate lobes; Drs. T. Antakly and M. Cantin for the immunocytochemistry; the secretarial help of Ms. D. Marcil is greatly appreciated.

REFERENCES

1. Li, C.H., in Hormonal Proteins and Peptides", (C.H. Li, ed.), Vol. 3, p. 1, Academic Press, New York.
2. Sairam, M.R., and Papkoff, H., in "Handbook of Physiology", (Nobil, E. and Sawyer, W.H., eds.) Vol. 4, p. 111, American Physiological Society, Washington, D.C. (1974).
3. Boileau, G., Seidah, N.G., and Chrétien, M., in "Hormonal Proteins and Peptides", (C.H. Li, ed.) Vol. 10, p. 65, Academic Press, N.Y. (1981).

4. Brownstein, M.J., Russell, J.T., and Gainer, H., Science 207, 373 (1980).
5. Seidah, N.G., Benjannet, S., and Chrétien, M., Biochem. Biophys. Res. Commun. 100, 901 (1981).
6. Watson, S.J., Seidah, N.G., and Chrétien, M., Science 217, 853, (1981).
7. Watson, S.J., Akil, H., Ghazarossian, V.E., and Goldstein, A., Proc. Natl. Acad. Sci. USA 78, 1260 (1981)
8. Steele, A.K., McCann, S.M., and Negro-Vilar, A., Endocrinology 111, 722 (1982).
9. De Palatis, L.R., Fiorindo, R.P., and Ho, R.H., Endocrinology 110, 282 (1982).
10. Johnson, L.K., Baxter, J.D., Vlodavsky, I., and Gospodarowicz, D., Proc. Natl. Acad. Sci. USA 77, 394 (1980)
11. Antakly, T. and Feigelson, P., J. Cell. Biol. 91, 231a (1981).
12. Antakly, T., Pelletier, G., and Feigelson, P., Proc. Natl. Acad. Sci. USA (In Press) (1983).
13. Hsi, K.L., Seidah, N.G., De Serres, G., and Chrétien, M. FEBS Lett. 147, 261 (1982).
14. Larivière, N., Seidah, N.G., and Chrétien, M., Int. J. Pept. Prot. Res. 18, 487 (1981).
15. O'Farrell, P.H., J. Biol. Chem. 250, 4007 (1975).
16. Seidah, N.G., Rochemont, J., Hamelin, J., Lis, M., and Chrétien, M., J. Biol. Chem. 256, 7977 (1981).
17. Lazure, C., Seidah, N.G., Chrétien, M., Pelletier, R., and St.Pierre, S., Can. J. Biochem. (In Press) (1983).
18. Antakly, T., Lynch, K.R., Nakhasi, H.L., and Feigelson, P., Amer. J. Anat. 165, 211 (1982).
19. Dayhoff, M.D., and Orcutt, B.C., Proc. Natl. Acad. Sci. USA 76, 2170 (1979).
20. Seizinger, B.R., Hollt, V., and Herz, A., Biochem. Biophys. Res. Commun. 103, 256 (1981).
21. Spampinato, S., and Goldstein, A., Neuropeptides 3, 193 (1983).
22. Duello, T.M., and Nett, T.M., Mol. Cell. Endocrinol. 19, 101 (1980).

PART III
Neuropeptide Deficiencies

MULTIPLE FORMS OF NEUROPEPTIDES IN HUMAN CSF AND THEIR SIGNIFICANCE

Lars Terenius
Fred Nyberg

Department of Pharmacology
University of Uppsala
Uppsala
Sweden

I. INTRODUCTION

Despite their large numbers and wide distribution in the CNS, the functional importance of neuropeptides remains elusive. This is partly caused by limitations of experimental approaches. Peptides and peptide analogues only cross the blood-brain barrier with difficulty if at all, invalidating studies where peptides are given systemically. Direct injection of peptides into the CNS or CSF spaces may cause effects which may, however, be artifactual due to this artificial way of adminstration. Pharmacological tools such as antagonists are rarely available and if available are seldom perfect. Agents capable of producing release, blockade of synthesis or degradation have so far been found to be fairly nonselective, which is not entirely unexpected because such mechanisms seem to be fairly general.

A different general approach is to study the release of an agent. Whereas peptide release in endocrine systems may be measured simply and fairly accurately by plasma analysis, neuropeptide release within the CNS is difficult to assess in general, and particularly so in patients. Our approach has been to analyze CSF in patients, to monitor activity in the two proenkephalin and substance P systems. Such studies have been directed to various clinical conditions including acute and chronic pain (1) and affective disorders (2) in view of the general neurotropic effects of opiates

BIOCHEMICAL AND CLINICAL ASPECTS OF NEUROPEPTIDES: SYNTHESIS, PROCESSING, AND GENE STRUCTURE

ISBN 0-12-417320-9

on mood and behaviour. The consequences of long-term opiate (heroin) abuse have also been investigated (3). There are several limitations to the use of CSF analysis. The fluid collected derives from the lumbar region of the spinal canal and its composition may not accurately reflect activity at higher CNS centres. Moreover, measured activity in CSF can only give an indication of the general level of activity in a particular peptide system and changes in smaller areas of the CNS will probably not be possible to observe. Finally, the wide-spread distribution of neuropeptides and the absence of clear-cut functional attributes to any one system will also make interpretations difficult. The complexity of the system under study makes it necessary to use an empirical approach. During this work we have observed problems which were not anticipated and relating to the existence in the CSF of multiple peptides other than those currently held to be functionally significant in the CNS. We have also come across the possibility that there may occur genetically determined differences in neuropeptide systems which may have pathogenetic consequences.

II. OPIOID PEPTIDES IN CSF - THE PROBLEM

In very early experiments which were initiated even before the structures of any endorphin was known, we found that opioid activity could be isolated from human CSF (4). Using a simple Sephadex G-25 column we could fractionate the active material into two major low-molecular weight fractions, which reacted in an opioid receptor assay (Fig. 1). As structures of endorphins isolated from brain, pituitary or adrenal became known it also became possible to synthetize these peptides, generate antibodies and develop specific radioimmunoassays for them. It then became apparent that opioid activity that was found to be present in CSF could not be accounted for simply by those peptides found to occur in large quantities in CNS or adrenal (Fig. 1). This raised several questions and suggested several possibilities. For instance quantitatively dominating peptides might be the best markers of release. More interestingly, the peptides identified in the CSF might have a functional importance by themselves and as an extreme alternative, might even be the major products of release in the synaptic area. It therefore seemed very important for us to identify the chemical structure of these CSF peptides. Since concentrations are low in the CSF and the access of fluid is extremely limited for ethical reasons this is not an easy task. Moreover, there is so much chemical homology between the various peptides that this causes large technical problems of separation.

The chemical complexity of opioid peptides became possible to systemize only by the results of genetic technology, identifying the precursors of the enkephalins, proenkephalin (5), of dynorphin, proenkephalin B (6) and of beta-endorphin, proopiomelanocortin (7). A whole series of opioid peptides would generate from these precursors, many of which had already been isolated and identified chemically (Table I). This structural information has been instrumental in our attempts to determine the generic relationships between peptides that we identify in CSF and the different precursors.

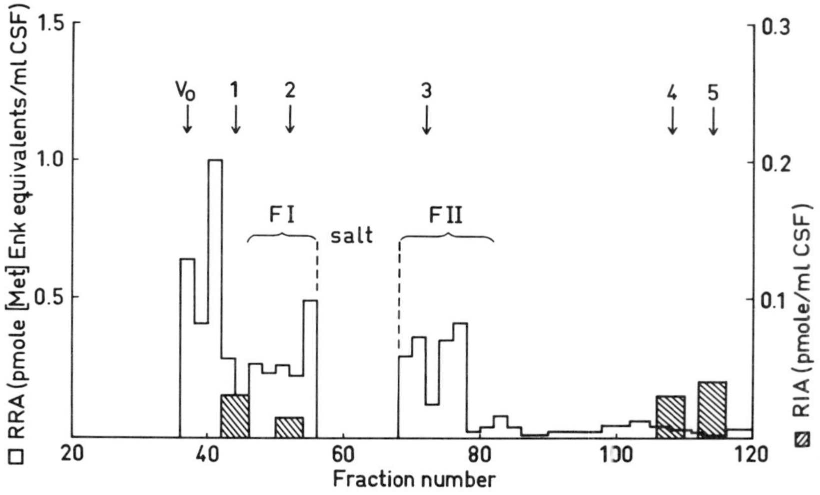

FIGURE 1. Sephadex G-10 chromatography of human lumbar CSF (50 ml) in 0.2M acetic acid. Fractions of 20 ml were collected at a flow rate of 1 ml/min. The column (50 cm x 100 cm) was calibrated with endorphin standards as indicated by the arrows (1=beta-endorphin; 2=dynorphin; 3=enkephalinyl-hexapeptides; 4=(Leu)enkephalin; 5=(Met)enkephalin. The radioreceptorassay was performed using opiate receptors of synaptic plasma membrane from rat brain and tritium-labelled dihydromorphine as competing radioligand (4). The binding activity is expressed in (Met)enkephalin equivalents which was used as a reference. Data obtained by radioimmunoassays for beta-endorphin, dynorphine, (Leu)- and (Met)enkephalin are also inserted in the figure. The radioimmunoassays were conducted as described elsewhere (12).

TABLE I. Examples of possible opioid peptides derived from processing of the respective precursors

Precursor	Peptide	No in precursor
Proenkephalin	Met-enkephalin	4
	Leu-enkephalin	1
	Met-enkephalinArg^{6}Phe^{7}	1
	Met-enkephalinArg^{6}Gly^{7}Leu^{8}	1
Proenkephalin B	Dynorphin	1
	Dynorphin B (Rimorphin)	1
	Alfa-neoendorphin	1
Proopiomelanocortin	Beta-endorphin	1

III. OPIOID PEPTIDES IN CSF, THE APPROACH

Peptides in CSF occur in very low concentrations. The original Sephadex procedure served very well in separating two larger peaks of opioid activity, Fractions I and II, with good recovery. High-molecular weight material which artifactually interferes in receptor assay was also removed by the procedure (Fig. 1). Preliminary studies indicated that the material in both fractions I and II was heterogenous and that further separation could be obtained by for instance high-performance liquid chromatography (8). Although this is a powerful method for peptide separation it was not ideal for our purposes; it required more extensive clean-up than achieved by the Sephadex separation step and recoveries tended to be quite low for certain components. It was therefore deemed necessary to introduce an extra separation step, which could accept material from the gel filtration column and provide good resolution. We decided to try electrophoresis in very dilute agarose gel which was found to separate the opioid peptides largely according to charge (9). It was then found that Fractions I and II separated into several components. Fig. 2 shows the results obtained using this technique on CSF samples from three different patients. Fraction I seems to separate into at least five different components, presumably differing in charge (cf migration rates of standards) and Fraction II into at least two components. These three patients had a spinal tap made in connection with a suspected increase in

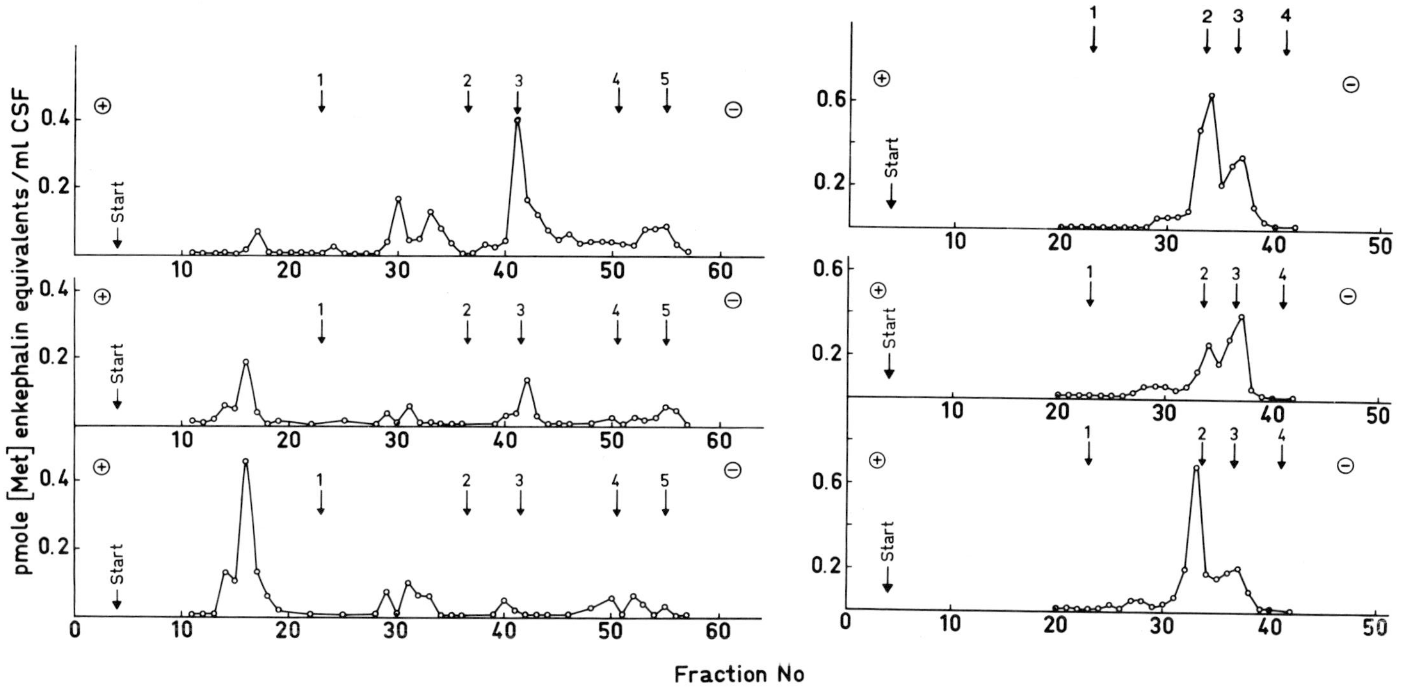

FIGURE 2. Column electrophoresis in 0.18% (w/v) agarose suspension of FI (left panel) and FII (right panel) material originating from CSF samples from three individual patients. Electrophoresis was performed in 0.10 M ammonium formate, pH 2.7, at a voltage of 1000V during 6 h. Opioid activity was monitored by a radioreceptorassay. The positions of some reference peptides are also shown left panel: 1=(Met)Enkephalin; 2=(Met)Enk-Lys^6; 3=dynorphin(1-8); 4=dynorphin(1-9); 5=dynorphin(1-13); right panel: 1=(Met)Enk; 2=(Met)Enk-Arg^6-Phe^7; 3=(Met)Enk-Lys^6; 4=dynorphin (1-8).

intracranial pressure due to obstructed CSF dynamics. In these three cases, no pathology was observed and there was no evidence for a blood-brain barrier defect. The individual differences may therefore reflect the "normal" variation in peptide patterns as we have also observed them in other cases. It is clear that under such circumstances, measurement of one of the many subfractions may not necessarily reflect the activity in a particular opioid peptide system. In fact, closer inspection of the relative proportions of the subfractions from Fraction I reveal a potential precursor product relationship, the peak migrating slower than enkephalin being inversely related in size to the peak migrating as dynorphin (1-8). Clearly, quantitation of all these peptides would require a major effort both in terms of structure identification and establishment of analytical procedures.

Further progress in structure determination is limited by the comparatively small volumes of CSF available. Definite structures of individual components by amino acid sequence determination or mass spectrometry may be possible in particular for small peptides fragments. In most instances structures will be tentatively assigned by comparison with reference components using various chromatographic procedures and alternative assay techniques such as specific radioimmunoassay. It should be stressed that this is the usual for material isolated from more easily accessible tissue sources such as brain; very few peptides identified in brain have actually been isolated and subjected to rigorous structure determination. Tentative structure assignments via electrophoresis (cf Fig. 2), HPLC and radioimmunoassay have been made for one component in Fraction II which is likely to be Met-enkephalin-Lys^{6} and one Fraction I component probably being dynorphin (1-8). In confirmation of earlier findings, levels of these two tentatively identified peptides are in at least tenfold excess over those of the enkephalins, beta-endorphin and dynorphin.

The use of separation procedures which provide some information about the chemical structures of the separated components such as shown here for electrophoresis, also provides a tool for demonstrating the existence of abnormal products of processing. The frequency of such abnormalities is not presently known and would require quite extensive studies. We have previously described a young child with a clinical syndrome suggestive of hypersecretion of endorphins (10). This child presented repeated attacks of apnea, which eventually became fatal at the age of 2.5 years, markedly reduced pain sensitivity and pin-point pupils, signs and symptoms characteristic of morphine poisoning. The attacks were reversed by naloxone. Analysis of CSF revealed extremely elevated levels of opioid activity in both Fractions I and II. Postmortem analysis of brain biopsy specimens showed very high levels of enkephalin as measured in radioimmunoassay. This child's younger brother showed essentially the same symptoms at the age of 7 months and he died a few months later. Also this child showed very high CSF endorphin levels. The rare incidence of this syndrome and its occurrence in two brothers strongly suggest a genetic linkage. Preliminary chemical characterization of CSF endorphins indicates that the major Fraction II endorphin is unrelated to those quantitatively dominating in "normal" human CSF. Similar indications of abnormal endorphin peptides have been demonstrated in patients with psychotic disorders (11).

Elucidation of the complexity of opioid peptide patterns in individual patients requires a multitude of procedures which is very demanding on CSF quantities. To meet this problem we have developed as an alternative to the Sephadex G25 method (Fig. 1) an ion-exchange procedure which gives group separation according to charge (12). The separation procedure is followed by a specific radioimmunoassay of the peptides listed (Table II). Thus, the same CSF sample can be used for several assays, which is essential for our sensitivity requirements. Another possibility which can address the generic origin of a particular opioid peptide at a different level is to use tryptic digestion and subsequent assay specifically for the tryptic fragments. The fragment Met-enkephalin-Lys^6 is unique to the proenkephalin precursor and Leu-enkephalin-Arg^6 to the proenkephalin B precursor. Both fragments occur in triplets in the respective precursors. Specific radioimmunoassay procedures have been developed for these two peptides and we are presently investigating a procedure of tryptic digestion followed by the specific radioimmunassays as a generally applicable assay procedure. The limitations of this enzymatic degradation approach is that certain peptide fragments such as those deriving from the proopiomelanocortin precursor will escape detection.

IV. CSF ENDORPHINS AS CLINICAL MARKERS

All the well-known profound CNS effects of opiates can be reproduced more or less completely by one or the other of the opioid peptides. If these peptides are released endogenously corresponding effects should be noticeable. We

TABLE II. Separation of neuropeptides by step-wise elution from a SP-Sephadex C-25 ion-exchanger. Three ml of CSF is run on a 1 ml disposable column

Elution step	Buffer Pyridine/Formic acid	pH	Peptide eluted
I	0.018M/0.1M	3.2	(Wash)
II	0.1M/0.1M	4.4	(Leu) enkephalin (Met) enkephalin
III	0.35M/0.35M	4.4	(Leu) enkephalin-Arg^6 (Met) enkephalin-Lys^6
IV	0.8M/0.8M	4.4	Substance P
"	1.6M/1.6M	4.4	Dynorphin

have conducted several clinical studies where Fraction I and Fraction II endorphins have been measured and correlations to various clinical variables have been sought. As an example, the results obtained in patients with chronic pain and affective disorders will be discussed. In chronic pain patients it was found empirically that levels of opioid activity in Fraction I were frequently very low. In an early study (4) this was believed to be a general feature of chronic pain and perhaps a sign of exhaustion. However, another study which was more balanced with regard to the representation of various types of chronic pain conditions revealed that the low Fraction I levels were characteristic of chronic neurogenic pain (Table III). In such such cases, pain originates within the nervous system and may be secondary to physical trauma or a virus infection. In chronic pain occurring in patients with cancer or ischemia, somatogenic pain, there were no abnormalities in the measured endorphin fractions. Finally, in chronic pain better explained in psychologic terms and where the patient has no sign of somatic lesions the endorphin Fraction I levels are above those in healthy volunteers and approaching those in affective disease. This latter finding suggests a nosological similarity, which in fact has been proposed earlier on psychiatric grounds. Thus, the measurements suggest a clinically meaningful classification of chronic pain. Perhaps more importantly is the independent finding that this classification has predictive power for the clinical response rate to transcutaneous nerve stimulation: After a three-month observation period, 90 patients with

TABLE III. Distribution of cases in different clinical categories with regard to Fraction I endorphin levels in CSF

Clinical diagnosis	Fraction I (pmol/ml CSF)*		
	<0.6	0.6-1.2	> 1.2
Healthy volunteers	3	12	4
Neurogenic pain	29	2	2
Other organic pain	2	3	3
Psychogenic pain	3	9	10
Unipolar depression		3	12

*Calculated as Met-enkephalin equivalents.

chronic neurogenic pain showed a 87% response rate, 16 patients with somatogenic pain 44% and 17 patients with psychogenic pain, only 24% (13). It is also known that at least with certain stimulation variables the therapeutic effect is naloxone-reversible and accompanied by increases in Fraction I endorphins. It then seems logical to assume a causal link between low endorphin levels and high response rate to electric stimulation in patients with chronic neurogenic pain. In etiological terms, these pain conditions may be more related to a lack of pain modulation than of excessive afferent influx.

A therapeutic effect, comparable to that obtained with transcutaneous nerve stimulation is rarely obtained with drugs. Not even morphine is very effective, besides all reservations one may have against this drug in a chronic condition. It therefore seems of importance to establish the mechanism of action of this and related procedures, such as acupuncture (14). Studies of this kind are therefore being carried out in this laboratory. In order to approach these mechanisms more closely we are also using an experimental model, the anaesthetized cat which is subjected to continuous spinal perfusion. In this model we can evaluate the effects of different kinds of afferent stimulation thus mimicking acupuncture or TNS. With the perfusion technique, substances released into the CSF can be rapidly recovered and are more likely to escape metabolic degradation than in lumbar CSF with its slow

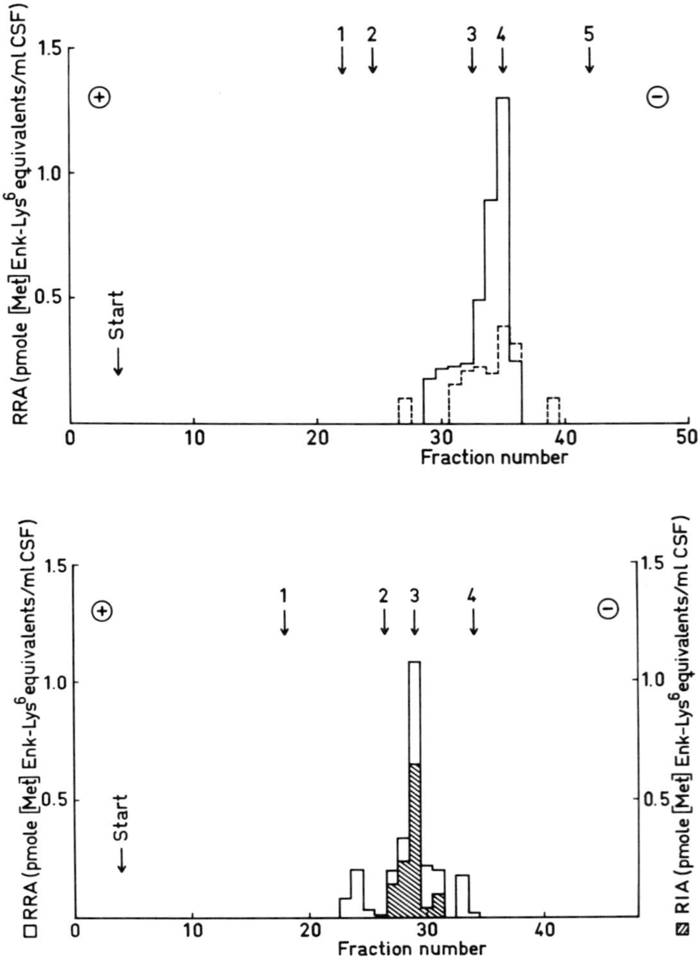

FIGURE 3. Electrophoretic separation of Fraction II activity in spinal perfusate from anaesthetized cats. Perfusion was carried out at resting conditions or during high intensity bilateral electric stimulation of the sciatic nerves.

Upper panel: Samples collected during resting (-----) and stimulation (———) at 1000V for 6 h. Fractions were tested in receptorassay. Lower panel: (Met)enkephalin-Lys^6 immunoactivity and opioid receptoractivity after 5 h electrophoresis of a sample collected during stimulation. The column was calibrated with (Upper panel): 1=(Met)Enk; 2=(Leu)Enk; 3=(Met)Enk-Arg^6-Phe^7; 4=(Met)Enk-Lys^6; 5=dynorphin(1-8); (Lower panel): 1=(Met)Enk; 2=(Met)Enk-Arg^{6}Phe^7; 3=(Met)Enk-Lys^6; 4=dynorphin (1-8). For other details, see legend to Fig. 2.

circulation and turnover. Furthermore, one can add protease inhibitors to the perfusate to inhibit the local breakdown. Using this protocol we can show that strong afferent stimulation causes massive increases in the efflux of opioid peptides (15). Whereas dynorphin or enkephalin contribute very moderately to this increase in terms of opiate receptor occupancy, there is a very marked increase in both Fractions I and II (Fig 3, upper panel) activity. The electrophoretic mobility of the activity suggests that the increase may be due to Met-enkephalin-Lys^{6} which is also confirmed by radioimmunoassay (Fig. 3, lower panel).

V. EXPERIENCES WITH MEASUREMENT OF SUBSTANCE P IN HUMAN CSF

Substance P fibers have a regional distribution within the CNS related to that of the enkephalins. This association may have functional significance. For instance, at the spinal level, the nonmyelinated fibers involved in pain conduction produce substance P. These fibers terminate in the dorsal spinal cord and the termination area receives input from enkephalin fibers. It has been postulated that substance P may be a pain transmitter and that some of the "gating" of pain signals at the spinal level originates from enkephalin interneurons (16). At higher CNS levels, there are also topological associations between enkephalin, sometimes dynorphin, and substance P fibers. This suggests that it may be meaningful to measure substance P in the CSF in parallel with endorphins.

Preliminary studies were conducted using the ion-exchange procedure outlined in Table II for preseparation prior to radioimmunoassay. Despite numerous attempts, sometimes involving 10-20 ml CSF, we were unable to identify substance P. Recoveries of added substance P were excellent and the peptide appeared stable. Since there were reports of substance P levels in human CSF in the literature (17) we tried other separation procedures and found immunoactive substance. This material also emerged from the ion-exchanger but at lower ionic strength, suggesting that the material was less polar. This was confirmed using our electrophoresis procedure. Levels of substance P were below detection limits whereas material migrating as the (3-11) and (5-11) fragments were detectable (Fig. 4). Using a radioimmunoassay specific for the C-terminus of substance P (1-7), we could also show activity, which, however, did not migrate identically with the reference compound (Fig. 4). Further work is necessary before the exact chemical nature of these radio-

TABLE IV. Receptor preferences of some opioid peptides (19).

Peptide	Receptor preference	
	Primary	Secondary
Enkephalins	δ	μ
(Met)enkephalinArg^{6}Phe^7	δ μ	
Dynorphin (1-8)	κ	μ
Dynorphin	κ	μ
Beta-endorphin	μ κ δ	(nonselective)

immunoactive fragments can be established. These finding have prompted us to develop procedures to separate and measure substance P and its fragments in brain tissue. We have so far observed large species differences and regional differences within species in levels of C-terminal fragments and SP(1-7) relative to substance P. With such information and at least some chemical separation prior to radioimmunoassay, we may generate more meaningful information from CSF analysis than if we were just measuring "substance P-like radioimmunoactivity".

VI. CONCLUDING REMARKS

Our studies on neuropeptides in CSF have given information at different levels. The potential multitude of peptides which can arise from a precursor is also reflected in a large number of peptides present. The peptides identified in CSF may be minor products from the processing of the precursor which survive metabolic degradation but they may also be important signal substances by themselves. An extreme possibility, which cannot be discarded, is that peptides present in the CSF are the functionally most important ones. It should also be kept in mind that for instance, C-terminal extension of enkephalins changes receptor selectivity (Table IV) and that substance P (1-7) has a different action profile in the CNS than the mother compound (18). Finally, it has been emphasized that our approach is potentially able to reveal abnormalities, maybe genetically linked, in peptide processing and degradation.

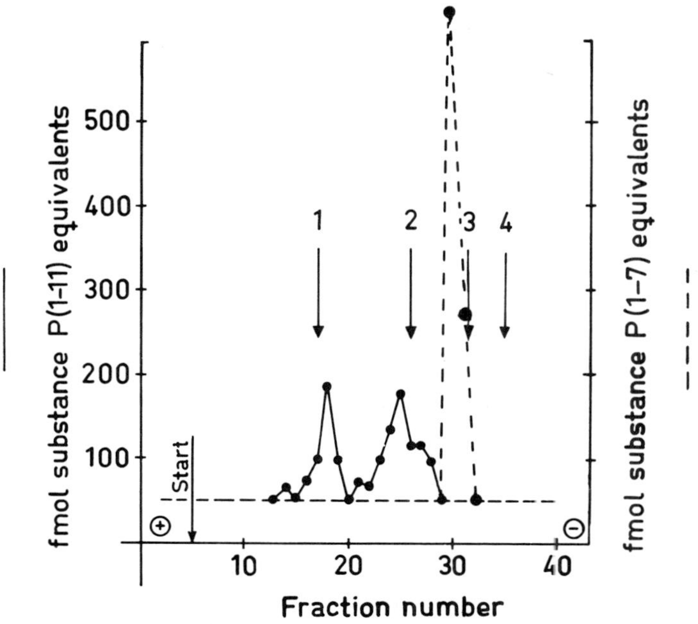

FIGURE 4. The distribution of substance P(1-7) and substance P(1-11) immunoactivity after electrophoretic separation at pH 2.7 of CSF material originating from a group of patients with unipolar depression. Electrophoresis was for 5 h at a voltage of 1000V. The column was calibrated with authentic substance P and its fragments (1=fragment (5-11); 2=fragment (3-11); 3= substance P; 4=fragment (1-7).

The radioimmunoassays have been described elsewhere (12). For further details, see legend to Fig 2.

ACKNOWLEDGEMENTS

This work is supported by the Swedish Medical Research Council and the National Institute on Drug Abuse, Washington D.C.

REFERENCES

1. Terenius, L., Front. Horm. Res., 8, 162 (1981).
2. Ågren, H., Terenius, L., Wahlström, A., Ann. N.Y. Acad. Sci., 398, 388 (1982).
3. O'Brien, C.P., Terenius, L., Wahlström, A., McLellan, A.T., Krivoy, W., Ann. N.Y. Acad. Sci., 398, 377 (1982).

4. Terenius, L., Wahlström, A., Life Sci., 16, 1759, (1975).
5. Gubler, U., Seeburg, P., Hoffman, B.J., Gage, L.P., Udenfriend, S., Nature, 295, 206 (1982).
6. Kakidani, H., Furutani, Y., Takahashi, H., Noda, M., Morimoto, Y., Hirose, T., Asai, M., Inayama, S., Nakanishi, S., Numa, S., Nature, 298, 245 (1982).
7. Nakanishi, S., Inoue, A., Kita, T., Nakamura, M., Chang, A.C.Y., Cohen, S.N., Numa, S., Nature, 278, 423 (1979).
8. Wahlström, A., Terenius, L., FEBS Lett., 118, 241 (1980).
9. Nyberg, F., Wahlström, A., Sjölund, B., Terenius, L., Brain Res., 259, 267 (1983).
10. Brandt, N.J., Terenius, L., Brock Jacobsen, B., Klinken, L., Nordius Å., Brandt, S., Blegvad, K., Yssing, M., N. Engl. J. Med., 303, 914 (1980).
11. Lindström, L., Nyberg, F., Terenius, L., Besev, G., Gunne, L-M., Lyrenäs, S., Lindberg, B., Am. J. Psychiat., submitted, 1983.
12. Bergström, L., Christensson, I., Folkesson, R., Stenström B., Terenius, L., Life Sci., submitted 1983.
13. Eriksson, M.B.E., Sjölund, B.H., Nielzen, S., Pain, 6, 335 (1979.
14. Han, J.S., Terenius, L., Ann. Rev. Pharmacol. Toxicol., 22, 193 (1982).
15. Yaksh, T., Terenius, L., Nyberg, F., Jhamandas, K., Wang, J-Y., Brain Res., in press, 1983.
16. Otsuka, M., Takahashi, T., Ann. Rev. Pharmacol. Toxicol., 17, 425 (1977).
17. Nutt, J.G., Mroz, E.A., Leeman S.E., Williams A.C., Engel, W.K., Chase T.N., Neurology 30, 1280 (1980).
18. Stewart, J.M., Hall, M.E., Harkins, J., Frederickson, R.C.A., Terenius, L., Hökfelt, T., Krivoy, W.A., Peptides, 3, 851 (1982).
19. Wüster, M., Rubini, P., Schultz, R., Life Sci., 29, 1219 (1981).

GENETIC LINKAGE ANALYSIS IN FAMILIAL DYSAUTONOMIA USING A DNA PROBE FOR THE β-NERVE GROWTH FACTOR GENE

Xandra O. Breakefield
Gregory Orloff
Carmela M. Castiglione

Department of Human Genetics
Yale University School of Medicine
New Haven, Connecticut

Felicia B. Axelrod

Department of Pediatrics
New York, University Medical Center
New York, New York

Lisa Coussens
Axel Ullrich

Genentech Incorporated
South San Francisco, California

A cloned DNA probe for the human β-nerve growth factor (β-NGF) gene has been used to exclude a role for this gene in the inherited neurologic disease, familial dysautonomia. Allelic contributions of the β-NGF gene were identified in genomic blots of parental DNA by variations in the length of homologous restriction fragments. The inheritance of the mutant gene causing dysautonomia and of specific β-NGF alleles was followed in families with two or more affected children. Linkage between these gene loci was excluded at a recombinational distance of 0 to 3 centiMorgans. This analysis demonstrates how DNA probes for neural genes can be used to exclude or implicate a role for these genes in inherited human diseases. Further, it resolves a long-standing question as to whether the structural gene for β-NGF is defective in dysautonomia.

This work was supported by NINCDS grant NS17083 and funds from the Dysautonomia Foundation and the McKnight Foundation to X.O.B. The linkage analyses were supported by NS11786 (to K. Kidd).

ISBN 0-12-417320-9

I. INTRODUCTION

With the exception of some of the gangliosidoses, essentially nothing is known about the molecular basis of inherited neurologic and psychiatric diseases in humans. Some of these diseases are undoubtedly caused by primary defects in proteins critical to the function of cells in the nervous system. In many inherited diseases, defects in the metabolism of other types of cells secondarily affect the nervous system. In this chapter only inherited defects in neural proteins will be considered.

The nervous system is in a unique genetic category in that it transcribes more of its genes than any other tissue; it has been estimated that over 50,000 genes are uniquely transcribed in brain (Brown and Greenwood, 1982; Chaudhari and Hahn, 1983). By comparison, only a few hundred neurally active proteins and peptides have been described. Thus, even if it were possible to isolate all known neural proteins from patients with inherited neuropsychiatric diseases and compare their structure with those from controls, less than one-hundredth of the possibly defective gene products would be examined. And, in fact, it is very difficult to obtain even the known neural proteins for study from humans.

Techniques of recombinant DNA technology provide two new approaches to the study of inherited neuropsychiatric diseases (See also Shows et al., 1982). Both approaches are most efficiently pursued if the disease being studied fulfills some of the following criteria: 1. it is genetically homogeneous (not caused by mutations at more than one gene locus); 2. it is inherited in a clear-cut Mendelian fashion (autosomal dominant, autosomal recessive, or X-linked); 3. it is fully penetrant (all individuals with the defective gene(s) express the disease); 4. it has an early age of onset, and can be readily diagnosed; and 5. it cannot be mimicked by non-genetic factors.

Many inherited neurologic diseases are appropriate for study, e.g. familial dysautonomia, neurofibromatosis, ataxia telangiectasia, Huntington disease, myotonic dystrophy and dystonia musculoram deformans. In one approach the chromosomal position of the sequence of DNA causing the disease is located by linkage analysis using common variations in DNA sequence spaced along the entire human genome. The methodology for this approach is being actively pursued in a number of laboratories and has been fully described elsewhere (Botstein et al., 1980; Housman et al., 1982). Once the position of the defective DNA sequence has been identified it should be possible eventually to determine what its normal function is, e.g. coding for a protein or regulating gene expression.

In the other approach, cloned DNA probes are obtained for genes coding for known neural proteins or peptides. Variations in DNA sequence in and around these genes are identified on the basis of variations in the length of restriction endonuclease fragments of genomic DNA that hybridize to the gene probe. These variations are used to follow the inheritance of alleles for a gene locus in families affected with an inherited neurologic disease to determine whether there is co-inheritance of the disease state with specific alleles (Breakefield et al., 1983a). This approach can be employed when a specific neural protein is a suspected candidate for the disease process, and a DNA probe coding for the structural gene in question is available. The advantage of this second approach is that it can be done on a relatively small number of nuclear families with two or more affected children, and information is obtained for coding and non-coding regions of the gene locus, as well as for substantial stretches of DNA sequence on either side of the locus.

Here we describe our studies using this second approach to establish whether the structural gene locus for β-NGF is defective in familial dysautonomia. This disease is inherited in an autosomal recessive manner with a high prevalence in the Ashkenazic-Jewish population suggesting that a single mutant gene is responsible for all classic cases (Riley et al., 1949; Moses et al., 1967; Axelrod, 1983) . Penetrance is complete and individuals are affected from birth. Symptoms include vomiting crises, cardiovascular instability, lack of overflow tears and taste buds, and insensitivity to pain and temperature. Most symptoms can be accounted for by the marked depletion of neurons in sensory, sympathetic and some parasympathetic ganglia; there is no apparent involvement of the central nervous system (Pearson and Pytel, 1978 a and b; Pearson et al., 1978). The neuropathologic findings are strongly reminiscent of changes seen in rodents treated in utero or postnatally with antibodies to β-NGF (Gorin and Johnson, 1980; Pearson et al., 1983; Aloe et al., 1980; Mayer et al., 1982), and this has led to the hypothesis that β-NGF or a β-NGF-like protein might be defective in these patients (Pearson et al., 1974). Attempts to resolve this by measuring the amount and biological activity of β-NGF-like proteins in peripheral tissues such as serum (Siggers et al., 1976) and cultured skin fibroblasts (Schwartz and Breakefield, 1981) have been frustrated by potential artifacts of the assay systems used and by the fact that authentic human β-NGF has not been characterized (Breakefield et al., 1983a).

The availability of cloned genomic probes for the human β-NGF gene (Ullrich et al., 1983) has allowed resolution of the role of this gene in familial dysautonomia. The nucleotide sequence of the human gene indicates that it codes for a β-NGF protein with >90% homology to β-NGF. There is one locus in the haploid human genome coding for β-NGF and it is located on chromosome one (Francke, Coussens, de Martinville, and Ullrich, submitted).

II. METHODS

The work described in this chapter has appeared in part in Breakefield et al. (1983a); and the methods are described therein. Briefly, Ashkenazic families were chosen for study that had two or more children affected with classic symptoms of dysautonomia. Lymphoblast lines were established from all members of the immediate families using Epstein-Barr virus, and genomic DNA was isolated from all lines. Parental DNA was cut successively with restriction endonucleases known to have sites in the β-NGF gene region and to generate a small number of fragments in the size range of 1 to 10 kb that would hybridize to the probes used. Probes consisted of cloned fragments of the human β-NGF gene. Restriction endonucleases were tested until one was found that yielded common variations in fragment length that could serve to mark different alleles at the β-NGF locus. Genomic DNA from all family members was then analyzed using this informative endonuclease to establish whether variations were inherited in a co-dominant manner and whether all affected children within a family inherited the same alleles at the β-NGF locus. The position of the endonuclease site variations was inferred from the predicted and observed sizes of hybridizing genomic fragments, and from the lack of variations observed using other endonucleases and probes specific to the same region.

III. RESULTS AND DISCUSSION

A. Variations in DNA sequence in the β-NGF gene locus

1. β-NGF gene structure. Ullrich et al. (1983) have determined the nucleotide sequence for one normal allele of the human β-NGF gene, including almost all of the 3' flanking

sequence and most or all of the 5' flanking sequence. [Provided there is not a very large intervening sequence (>6.0 kb) at the 5' end (ibid).] Since the mRNA for human β-NGF has not been characterized, the structure of the human gene can only be inferred from the structure of the mouse cDNA (Ullrich et al., 1983; Scott et al., 1983). The human β-NGF gene and the probe used in this study are shown diagramatically in Fig. 1. The gene sequence codes for a larger precursor protein that is presumably processed to yield active β-NGF.

2. Restriction endonuclease digestion of genomic DNA. Genomic DNA from lymphoblasts derived from the parents and two grandparents of dysautonomic patients were digested to completion with a number of restriction endonucleases known to have sites in the β-NGF gene region, and hybridized to a ^{32}P-labeled probe for the β-NGF gene (Table 1). Out of ten endonucleases tested, two revealed variations in restriction fragment length among these parents. Hinc II yielded a 6.2 kb fragment in all cases except the father in family F for whom a 8.5 kb fragment was also observed (Fig. 4). Bgl II revealed three patterns of fragments in the parents. Three individuals (EI1, EI2, FII1) had both a 1.4 kb and 4.1 kb fragment; one

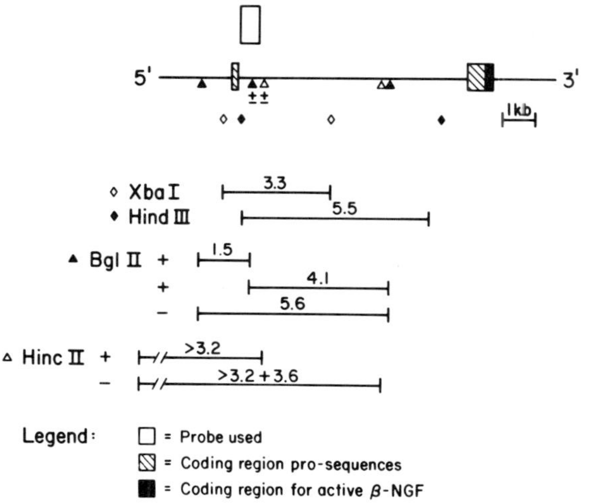

Fig. 1. Diagram of human β-NGF gene. The corresponding position and predicted sizes for fragments of genomic DNA cut to completion with Xba I, Hind III, Bgl II and Hinc II and hybridized to the probe (open box) are shown. Designations of + and - are given for fragments predicted for the presence or absence of a variable site. The symbol > is used to indicate the size of a fragment one end of which lies outside the known nucleotide sequence.

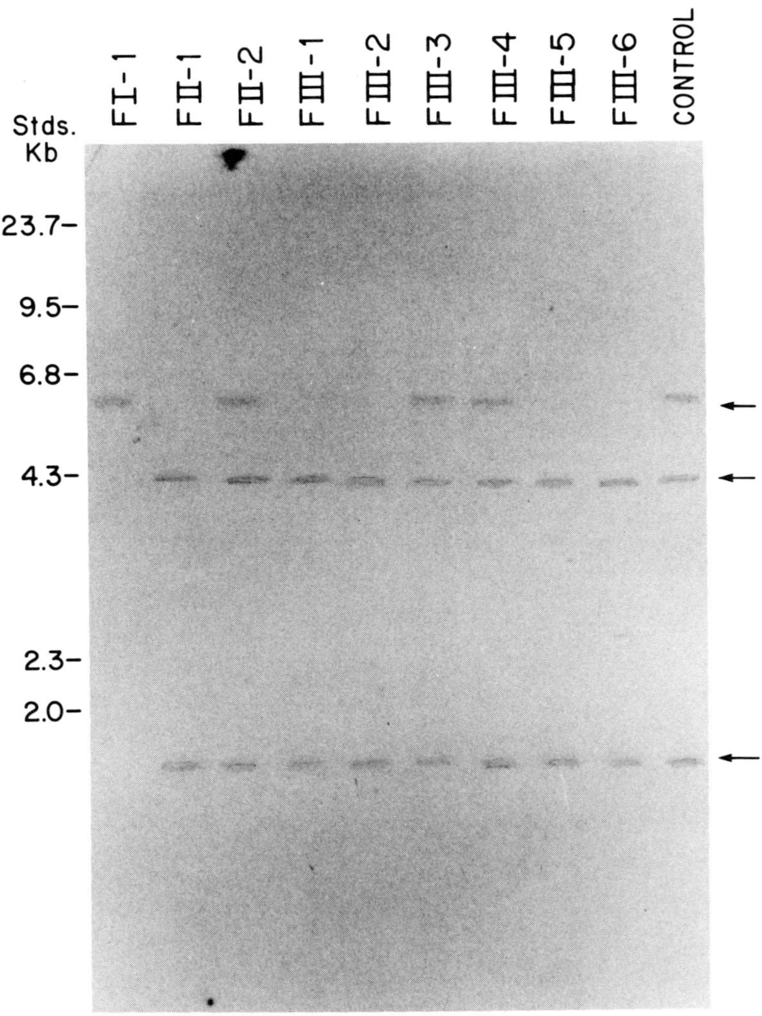

Fig. 2, Genomic Blot of DNA cut with Bgl II and hybridized to the β-NGF gene probe. DNA from family F was cut to completion with Bgl II and fragments resolved on a 1.0% agarose gel. Following denaturation and blotting to nitrocellulose, fragments were hybridized to a ^{32}P-labeled probe (Fig. 1). MW markers are derived from the position of fragments of Hind III digested λ DNA visualized with ethidium bromide. Hybridizing fragments of genomic DNA are shown by arrows. (Photo retouched).

(FI1), had only a 6.0 kb fragment; while nine out of the 13 parents and grandparents and one non-Ashkenazic-Jewish control showed all three fragments (Fig. 4). Bgl II digestions were performed on all members of the six families, e.g. Family F (Fig. 2), while Hinc II digestions only on members of Family F (Fig. 3). This extent of allelic variation was sufficient to exclude a role for the β-NGF gene locus in dysautonomia, provided that a defect in the same locus is responsible for all cases studied (see below).

3. Nature of variation in DNA sequence.
It is not possible to predict the exact nature of the sequence variations giving rise to fragment length differences observed in these families. The difference revealed by Bgl II appears to result from a small change in sequence including the variable (±) site indicated in the intervening sequence (Fig. 1). An insertion or deletion in this region would have

Table 1. Restriction endonuclease fragments hybridizing to a DNA probe for the β-NGF gene

Restriction endonuclease	Recognition site	Fragments	
		Predicted	Observed
		(kb)	
EcoRI	GAATTC	>2.9	6.7
Hind III	AAGCTT	5.5	5.6
Pst I	CTGCAG	3.1	3.1
Xba I	TCTAGA	3.3	3.5
Pvu II	CAGCTG	1.7	1.2
Sac I	GAGCTC	2.0	1.8
Stu I	AGGCCT	2.3	1.9
Hae II	RGCGCY	>4.0	4.9
Bgl II	AGATCT	1.5	1.4
		4.1	4.1
			6.0
Hinc II	GTYRAC	>3.2	6.2
			8.5

Genomic DNA from 12 parents was digested to completion with these restriction endonucleases. Fragment sizes were predicted from the known nucleotide sequence of the β-NGF gene (Ullrich et al., 1983); in cases where one of the cuts lies outside the known sequence > is used. Blots were hybridized to the probe shown in Fig. 1. Observed fragments were sized on the basis of migration of Hind III digested λ DNA in the same gel stained with ethidium bromide.

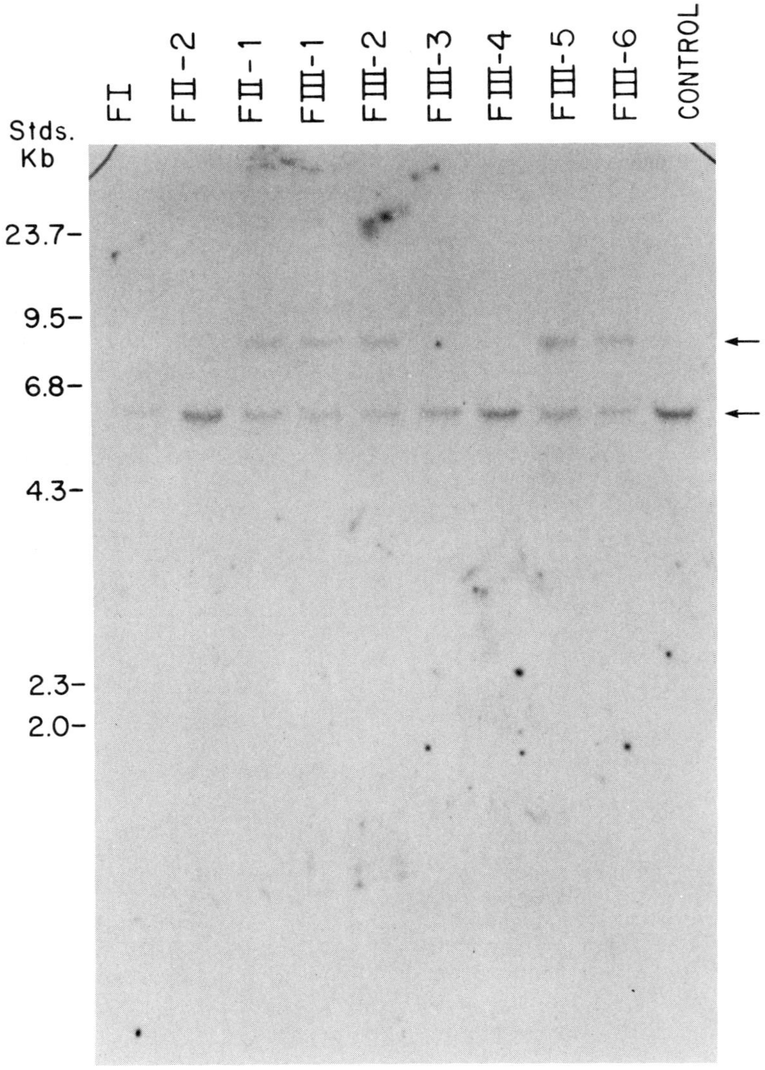

Fig. 3. Genomic blot of DNA cut with Hinc II and hybridized to the β-NGF gene probe. As in Fig. 2, DNA from family F was cut to completion with Hinc II; and a 1.0% agarose gel was used.

to be less than 0.1 kb in length, i.e. below the level of resolution in our gel system; as cutting with other endonucleases that generate a hybridizing fragment containing the variable Bgl II site did not reveal variable fragment lengths

(Fig. 1). Further, the observed fragment size 6.0 kb, is similar to the predicted one, 5.6 kb, assuming this site was lost by a single base-pair substitution within it. The position and nature of the variable Hinc II site is harder to resolve, as one end of the hybridizing fragment lies outside of the known sequence. If the variable site is in the intervening sequence (Fig. 1) a 3.6 kb difference between hybridizing fragments should have been observed in the heterozygous father, but a 2.3 kb difference was actually observed. This discrepancy cannot be explained by a 1.3 kb deletion including this site as it is not consistent with observations using other endonucleases that cut around this region (Xba I, Hind III). Alternately, the variable Hinc II site may lie in an unsequenced region at the 5' end of the hybridizing fragments.

On the basis of the number of sites tested using these ten endonucleases and a single probe for the β-NGF gene, and the number of nucleotides that must be in the correct sequence for cutting by each of them (Table 1), a discontinuous sequence in the β-NGF region of about 200 nucleotides has been compared among 13 individuals, or 26 alleles (see also, Breakefield et al., 1983b). Two variations in sequence were observed, both of which occur in non-coding regions of the gene and appear to be normal. This latter conclusion is drawn from the fact that the presence or absence of these sites varies independently from the carrier status of the parents for dysautonomia.

B. Inheritance of β-NGF alleles and dysautonomia

Both the Bgl II and the Hinc II variations in fragment length followed a pattern of co-dominant inheritance in these families (Fig. 4), as expected (Wyman and White, 1980). Further, in the case of family F, allelic haplotypes could be assigned to the father. The presence of the Bgl II site and the absence of the Hinc II site were co-inherited as a single allele, as were the absence and presence of these sites, respectively.

Families A, B, C and D are informative for the role of the β-NGF gene in dysautonomia, as at least one parent is heterozygous for the presence and absence of the variable Bgl II site and at least two children have dysautonomia. This latter point is critical as non-affected children may be either homozygous normal or heterozygous normal/affected at the disease locus. In families A and B, affected children may have inherited the same alleles at the β-NGF locus from their parents. In families C and D, however, affected children definitely did not inherit the same alleles at this locus. Thus, it is apparent that the β-NGF gene does not cause dysautonomia, since in an autosomal recessive disease, affected children must have the same alleles at the locus

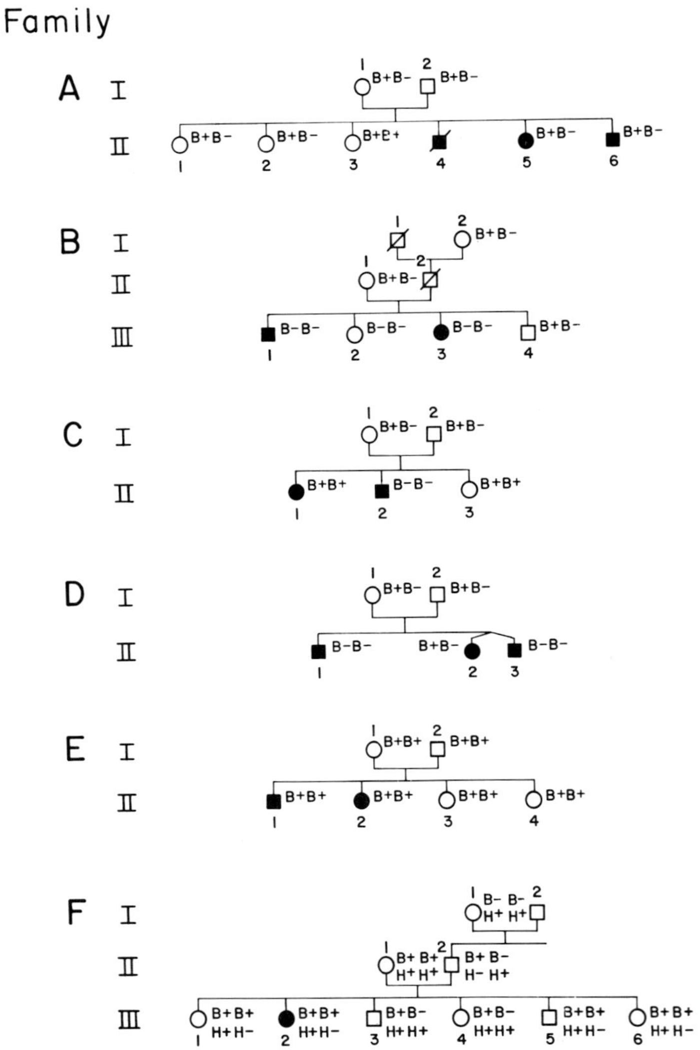

Fig. 4. Dysautonomia families showing segregation of variable site for Bgl II and Hinc II in the β-NGF gene. Affected individuals are shown as closed symbols. The alleles at the β-NGF gene locus are designated as B^+, B^-, H^+, and H^- according to fragment lengths observed in genomic blots from these family members. All analyses were carried out two-three times for each individual. This figure is reprinted from Breakefield et al. (1983b).

causing the disease. Linkage analysis of the data from these four families used the computer program LIPED (Ott, 1976).

The lod score (Z) was negative at all values of the recombination fraction (Θ), arguing against any linkage. The lod score was significantly negative ($Z<-2.0$) for all $\Theta \leq .03$. In other words, the odds are at least 100:1 against any recombination fraction less than 3%. Since a recombination frequency of $\Theta=.01$ (1 centiMorgan or map unit) has been crudely estimated as corresponding to 10^6 bp of DNA on average, this statistical exclusion of recombination within 3 centiMorgans excludes the gene for familial disautonomia from being within $3x10^6$ bp (3,000 kb) on either side of the gene for β-NGF (Fig. 5).

Even if all four informative families had shown apparent co-inheritance of the same alleles for the β-NGF gene in affected children, this would not have been sufficient proof for the role for this gene in dysautonomia. Rather it would have been necessary to continue testing other endonucleases to find additional fragment length variations that could mark more of the allelic contributions from parents in these families. At the rate the first two informative sites were observed about 20 more endonucleases with cutting sites in this gene region would need to be tested. If it were possible to distinctively mark all four allelic contributions for a gene from both parents in a family, all children would become informative. That is, not only should affected children in a family share the same alleles at the disease locus, but unaffected children should never have both these same alleles. If a pattern of inheritance were then observed consistent with that of linkage in these same four families, the gene being tested could be implicated with a LOD score of 4, that is the odds would be 10^4:1 favoring very close linkage ($\Theta\sim0.0$) over random segregation ($\Theta=.5$). However, significant deviation from independent segregation (i.e., a lod score $Z\geq+3.0$) would also be very likely if the DNA sequence causing the disease were within 1 centiMorgan (roughly 10^6 bp) of the gene being probed. Such a distance could easily encompass a cluster of genes in a gene family.

The power of this type of analysis lies in the speed with which it can be done and by the fact that it excludes or implicates any change in DNA sequence within or near a gene region, whether it is in a coding sequence or not. This is especially useful given that there is a large amount of normal variation in DNA sequence among genes, making it difficult to interpret the significance of sequence differences between genes from controls and patients (Prochownik _et al._, 1983; Phillips _et al._, 1982; Jeffreys, 1979; Higgs _et al._, 1981; Orkin _et al._, 1982; Ullrich _et al._, 1982; Bell _et al._, 1981; Rees _et al._, 1983)

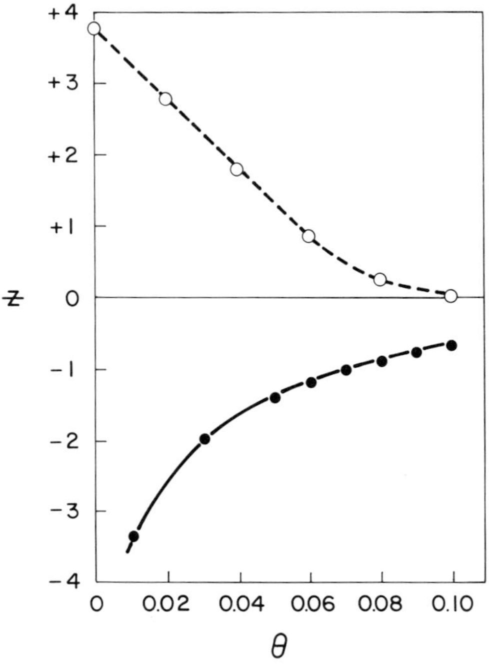

Fig. 5. LOD score analysis of segregation of dysautonomia and alleles at the β-NGF gene locus. Values were calculated using the computer program of Ott (1976). Closed circles show actual numbers calculated using data from Families A, B, C, and D. Open circles show theoretical numbers assuming data consistent with linkage in these families. Z = lod (log odds); θ = recombination fraction.

C. Prospectus

The study described here shows how techniques of molecular biology can be used to obtain a "yes" or "no" answer about the role of a particular neural protein in an inherited neurologic disease. The example of β-NGF in dysautonomia is a particularly striking one as this issue has been unresolvable by biochemical and immunologic techniques, and authentic human β-NGF has not yet been obtained and characterized. The use of cloned probes for neural genes expands the number of suspected proteins that can be investigated in inherited diseases with primary effect, on the nervous system. Fortunately, DNA probes generated for genes from other animals have substantial homology with and can be used to access human genes. A number of cDNAs for neural proteins have been or are being prepared, these include those for tyrosine hydroxylase (Lamouroux et al., 1982), phenylethanolamine-N-methyltransferase (Kaplan et al., 1983), dopamine β-hydroxylase (O'Malley et al., 1983), the nicotinic acetylcholine receptor (e.g. Ballivet et al., 1982) and several neuropeptides (e.g. Herbert et al., 1981; Land et al., 1982). DNA probes for many other neural proteins can be cloned at the current level of expertise available for them.

The exclusion of the structural gene for β-NGF in dysautonomia does not eliminate a role for defective β-NGF action in this disease. Action could also be disrupted through other proteins needed for β-NGF synthesis, storage and action including processing enzymes, the α and γ subunits of the NGF complex, and β-NGF receptors. Of these the last two are becoming imminently clonable as the amino acid sequence of the mouse γ subunit is known (Thomas et al., 1981), and human β-NGF receptors have been recently purified and antibodies prepared against them (Puma et al., 1983).

Now that a "permanent" genetic record of these dysautonomic pedigrees has been established through lymphoblast lines, it will be possible to test DNA from these individuals for co-inheritance of alleles for other NGF related proteins with the disease state. The same strategy can be employed as described here. Different endonucleases are tested until one or more is found to yield differences in the length of hybridizing fragments among parents. Affected children within a family are then examined for the presence of "similar" or "different" alleles. As observed here, it takes very little effort to exclude a gene as causing the disease. Implicating a gene requires much more work, but yields a clear answer that might not be apparent from a comparison of the nucleotide sequence of a few genes from patients and controls. This is especially true for sequence differences in intervening and flanking regions where variations can cause reduced rates of transcription and/or the formation of abnormal proteins (Orkin et al., 1982). An interesting feature of this type of analysis is that DNA sequences near the gene under study can also be excluded or implicated in the disease state. This may be especially useful in cases of gene families situated near each other on the genome. Related genes with substantial sequence homology can also be discovered using a probe for one of the family members, whether or not they are near each other in the genome, by screening genomic blots or libraries under conditions of reduced stringency (Scheller et al., 1983).

In summary, the increasing availability of DNA probes for genes coding for proteins important in the function of the nervous system will allow investigation of their role in inherited neuropsychiatric diseases. This approach will elucidate any kind of variation in DNA sequence in the structural gene that interferes with the production of normal gene product. Further, it provides a means of discovering new neural proteins that have structural similarities to known ones and of investigating their role in these diseases. This approach will be especially useful in the interval before a complete linkage map of the human genome is available and the genes coding for known proteins are positioned in it. Even after this, there will still be a large number of unmapped genes coding for unknown neural proteins. Many of these new genes will be found and mapped, in turn, through the fact that variation in their sequence can lead to inherited neuropsychiatric diseases.

ACKNOWLEDGMENTS

We thank the families who donated blood samples, and their physicians - Drs. Ralph Cash, Kenneth Keer, James Selwyn, John Graham and Leonard Pinsky, who obtained samples for us; Kenneth Kidd, James Gusella, Lois Morton and Susan Kruger for advice on linkage analysis; and Berengére de Martinville, Judith Kidd, James Gusella and Mary Anne Anderson for help with techniques.

REFERENCES

Aloe, L., Cozzari, C., Calissano, P., and Levi-Montalcini, R., Nature 291; 413 (1981).

Axelrod, F.B., in: "Cellular and Molecular Biology of Neuronal Development" (I.B. Black, ed.) in press, Plenum Press, New York (1983).

Ballivet, M., Patrick, J., Lee, J., and Heineman, S., Proc. Natl. Acad. Sci. USA 79, 4466 (1982).

Bell, G.J., Karam, J.H., and Rutter, W.J., Proc. Natl. Acad. Sci. USA 1981, 78:5759-5763.

Botstein, D., White, R.L., Skolnick, M., and Davis, R.W., Am. J. Hum. Genet. 1980, 32:314-331.

Breakefield, X.O., Castiglione, C.M., Coussens, L.M., Axelrod, F.B., Ullrich, A. in: "Cellular and Molecular Biology of Neuronal Development" (I. B. Black, ed.) in press, Plenum Press, New York (1983a).

Breakefield, X.O., Orloff, G., Castiglione, C.H., Coussens, L., Axelrod, F.B., and Ullrich, A., New Engl. J. Med., submitted (1983b).

Brown, I.R., and Greenwood, P., in "Molecular Approaches to Neurobiology" (I.R. Brown, ed.) p 41. Academic Press, New York (1982).

Chandhari, N., and Hahn, W.E., Science 1983, 220: 924-928.

Gorin, P.D., and Johnson, E.M., Jr., Devel. Biol. 1980, 80:313-323.

Herbert, E., Birnberg, N. Lissitsky, J.C., Civelli, O., and Uhler, M. Neurosci. Comment. 1, 16 (1981).

Higgs, D.R., Goodbourn, S.E.Y., Wainscoat, J.S., Clegg, J.B., and Weatherall, D.J., Nucleic Acids Res. 1981, 9:4213-4224.

Housman, D., Kidd, K., and Gusella, J.F., Trends in Neurosci. 1982, 5:320-323.

Jeffreys, A.J., Cell 1979, 18:1-10.

Kaplan, B., Jioio, A.E., and Batter, D.K., in "Gene Expression in Brain" (C. Zomzely-Neurath, ed.), in press. Wiley Interscience, New York, (1983).

Lamouroux, A., Biguet, N.F., Samolyk, D., Privat, A., Salomon, J.C., Pujol, J.F., and Mallet, J. Proc. Natl, Acad. Sci. USA 79, 3881 (1982).

Land, H., Schütz, G., Schmale, H., and Richter, D. Nature 295, 299 (1982).

Mayer, N., Lembeck, F., Geodert, M., and Otten, U., Neurosci. Lett. 1982, 29:47-52.

Moses, S.W., Roteni, Y., Jagoda, N., Talmor, M., Eichhorn, F., and Levin, S., Israel J. Med. Sci., 1967, 3:358-371.

O°Malley, K.L., Mauron, A., Raese, J., Barchas, J.D., and Kedes, L., Proc. Natl. Acad. Sci. USA 80, 2161 (1983).

Orkin, S.H., Kazazian, H.H., Jr., Antonarakis, S.E., Goff, S.C., Boehm, C.D., Sexton, J.P., Waber, P.G., and Giardina, P.J.V., Nature 1982, 296:627-631.

Ott, J., Am. J. Hum. Genet. 1976, 28:528-529.

Pearson, J., Axelrod, F.B., and Dancis, J., Ann. N.Y. Acad. Sci. 1974, 228:288-300.

Pearson, J., Johnson, E.M., and Brandeis, L. Devel. Biol., 1983, 96:32-36.

Pearson, J., and Pytel, B., J. Neurol. Sci. 1978a, 39:123-130.

Pearson, J., and Pytel, B., J. Neurol. Sci. 1978b, 39:47-59.

Pearson, J., Pytel, B., Grover-Johnson, N., Axelrod, F., and Dancis, J, J. Neurol. Sci. 1978, 35:77-92.

Phillips, J.A. III, Parks, J.S., Hjelle, B.L., Herd, J.E., Plotnick, L.P., Migeon, C.J., and Seeburg, P.H. J. Clin. Invest. 70, 489, (1982).

Prochownik, E.V., Antonarakis, S., Bauer, K.A., Rosenberg, R.D., Fearon, E.R., and Orkin, S.H., New Engl. J. Med. 308, 1549 (1983).

Puma, P., Buxser, S.E., Watson, L., Kelleher, D.J., and Johnson, G.L., J. Biol. Chem. 258, 3370 (1983).
Rees, A., Shoulders, C.C., Stocks, J., Galton, D.J., and Baralle, F.E., Lancet, 1983, 1:444-446.
Riley, C.M., Day, R.L., Greely, D. McL., and Langford, W.S., Pediatrics, 1949, 3:468-477.
Scheller, R.H., Jackson, J.F., McAllister, L.B., Rothman, B.S., Mayeri, E., and Axel, R., Cell 32, 7 (1983).
Schwartz, J.P., and Breakefield, X.O., Proc. Natl. Acad. Sci. USA 1980, 77:1154-1158.
Scott, J., Selby, M., Urdea, M., Quiroga, M., Bell, G.I., and Rutter, W.J., Nature, 1983, 302:538-540.
Shows, T.B., Sakaguchi, A.Y. and Nayler, S.L., Adv. Hum. Gen. 12, 341 (1982).
Siggers, D.C., Rogers, J.G., Boyer, S.H., Margolet, L., Dorkin, H., Banerjee, S.P.,and Shooter, E.M., New Engl. J. Med. 1976, 295:629-634.
Thomas, K.A., Baglan, N.C., and Bradshaw, R.A., J. Biol. Chem. 256, 9156 (1981).
Ullrich, A., Dull, T.J., Gray, A., Philips, J.A., and Peter, S., Nucl. Acid Res. 1982, 10:2225-2240.
Ullrich, A., Gray, A., Berman, C. and Dull, T.J., Nature, 1983, 303: 821-825.
Wyman, A.R., and White, R., Proc. Natl. Acad. Sci. USA 1980, 77:6754-6758.

CORRECTION OF GENETIC GnRH (GONADOTROPIN-RELEASING HORMONE) DEFICIENCY IN THE MOUSE BY PREOPTIC AREA TRANSPLANTS

Dorothy T. Krieger
Marie J. Gibson

Division of Endocrinology
Mount Sinai School of Medicine
New York, New York

Harry M. Charlton

Department of Anatomy
Oxford University
Oxford, United Kingdom

Mark J. Perlow

Department of Neurology
University of Illinois School of Medicine
West Side Veterans Administration Hospital
Chicago, Illinois

Earl A. Zimmerman

Department of Neurology
Columbia Presbyterian Hospital
New York, New York

Supported in part by grants from the Lita Annenberg Hazen Fund, NIH grant HD-05077, and the MRC.

ISBN 0-12-417320-9

INTRODUCTION

It has been demonstrated that transplants of portions of embryonic mammalian central nervous system (CNS) survive, differentiate, and become anatomically integrated with the CNS of neonatal and adult mammalian recipients (Azmitia *et al.*, 1981; Bjorklund *et al.*, 1979; Bjorklund and Stenevi, 1979; Das, 1974; Hallas *et al.*, 1980; Jaeger and Lund, 1980a; Jaeger and Lund, 1980b: Kromer *et al.*, 1981; Kromer *et al.*, 1979; Low *et al.*, 1981; Lund and Harvey, 1981; Lund and Hauschka, 1976; McLoon *et al.*, 1981; Oblinger *et al.*, 1980; Stenevi *et al.*, 1980a; Wells and McAllister, 1982). The transplants appear to have potential for promoting axonal regeneration (Bjorklund *et al.*, 1981; Bjorklund and Stenevi, 1981; Kromer *et el.*, 1981; Olson *et al.*, 1980a; Olson *et al.*, 1980b) and reducing functional deficits in the host brain caused by either lesions or congenital anomalies (Bjorklund *et al.*, 1981; Bjorklund *et al.*, 1980; Dunnett *et al.*, 1981; Freed *et al.*, 1979; Gash *et al.*, 1980; Krieger *et al.*, 1982; Kromer *et al.*, 1981; Low *et al.*,1981; Perlow *et al.*, 1979; Segal *et al.*, 1981). These cited studies have been concerned with exploring two major questions: (a) factors involved in the attainment of specific neural connections in the course of brain development; and (b) factors involved in the transplantation of neural tissue to correct surgically or genetically induced CNS defects. Our interests have centered on the latter. To this end, we have utilized the observations that (a) the hypogonadism of the hypogonadal mutant mouse (hpg) is secondary to a deficiency of hypothalamic GnRH and (b) in the normal animal the preoptic area is the site of GnRH-containing neurons, to investigate whether such a GnRH deficiency in the hpg mutant animal can be corrected by transplantation of preoptic area from unaffected animals of the hpg strain.

To more fully understand the nature of the results obtained, it will be helpful first to briefly review the normal regulation of the hypothalamic-pituitary-gonadal axis, the details of the nature of the gonadal deficiency of the hpg animal, and factors involved in functional transplants of CNS tissue.

Factors Involved in the Regulation of the Hypothalamic-Hypophyseal-Gonadal Axis

The principal locations of GnRH cell bodies in the rodent are within the septal and medial preoptic nuclei (Merchenthaler *et al.*, 1980), although GnRH neurons have also been reported to be present in the olfactory bulb (Jennes and Stumpf,

1980; Witkins *et al.*, 1982). The GnRH-containing fibers originating from all of these cells are widely dispersed in the CNS. The main GnRH pathway concerned with gonadotropin secretion courses through the mediobasal hypothalamus to terminate in the external layer of the median eminence. Such GnRH terminals are apposed to capillaries of the hypothalamic-hypophyseal portal system, by which route GnRH reaches the anterior pituitary. It has been shown that electrochemical stimulation of the POA results in proportional increases of GnRH in hypophyseal portal blood and of pituitary luteinizing hormone (LH) release (Eskay *et al.*, 1977; Turgeon and Barraclough, 1973). It is believed that under physiological circumstances release of GnRH occurs in a pulsatile fashion, concomitant with the ultradian LH pulses (Carmel *et al.*, 1976) which have been observed in a variety of species, and that such pulsatile release is essential for normal gonadotropin secretion.

In addition to the dense innervation of the median eminence, GnRH-positive fibers are found innervating other circumventricular organs, cingulate and neocortex, striae terminalis, medial amygdalar nucleus, habenula, and periaqueductal gray. Some of these pathways may be involved in the neural regulation of sexual behavior, for which an intrahypothalamic role of GnRH in facilitating steroid effects on such behavior has been suggested (Foremann and Moss, 1977; Pfaff, 1973; Sakuma and Pfaff, 1980).

With regard to neurotransmitter regulation of GnRH, major emphasis has been placed on the adrenergic system. The role of dopamine is controversial, with reports of both a stimulatory and an inhibitory role; such discrepancies may be secondary to variations in the hormonal milieu at time of study or in doses employed, which may differentially affect subclasses of dopamine receptors. Although serotonin, histamine, and acetylcholine have also been implicated, additional studies are required to clarify the nature of such control.

The central tegmental pathway contributes noradrenergic innervation to the POA (in addition to the supraoptic and arcuate nuclei), as well as to the internal layer of the median eminence, where axons terminate in close proximity to the above described GnRH terminals (Palkovits, 1981). These observations, together with physiological studies of catecholamine turnover (Rance *et al.*, 1981), responses to drugs affecting catecholamine synthesis and turnover or adrenergic receptors (Kalra *et al.*, 1980; Ratner, 1971; Sawyer *et al.*, 1947), and lesion experiments, have led to the hypothesis that catecholamines control or modulate tonic pulsatile and cyclic release of GnRH into portal blood, as well as mediating those aspects of the negative feedback effect of gonadal steroids which are effected via a hypothalamic locus of action. The role of such afferents to GnRH cells in the ontogenetic commitment of these cells to release GnRH remains a theoretical consideration. It has been observed that the Brattleboro rat, characterized by genetic absence of vasopressin, also has fewer fluorescent catecholamine varicosities in apposition to the vasopressin-deficient perikarya (Scholer and Sladek, 1981).

Nature of Hpg Animal and Its Neuroendocrine Status

The hypogonadal (hpg) mutant mouse appeared in a breeding colony at Harwell, England. The mutant mice of both sexes fail to show any sign of postnatal gonadal or accessory tissue development. However, sexual differentiation proceeds normally up to birth. (Both male and female mutants appear to possess normal olfactory function.)

Hormone assays and immunoperoxidase staining indicate that the primary cause of the hypogonadism is due to a severe deficiency in the hypothalamic releasing hormone (GnRH) which regulates pituitary luteinizing hormone (LH) and follicle-stimulating hormone (FSH) synthesis and release. (It is not known whether this deficiency is a result of gene deletion, failure of gene expression, or secondary to synthesis of an abnormal GnRH.) Such GnRH deficiency leads to the marked reduction in pituitary gonadotropin content present in mutants of both sexes, compared to normal littermates. In adult hpg mutants, ovarian follicles rarely reach the antral stage in the female, while in the male, spermatogenesis rarely advances beyond the diplotene stage.

Further evidence for a primary deficiency in GnRH comes from the observations that there is evidence of normal pituitary and gonadal function in the hpg mutant in response to stimuli normally effective at these levels. Although in the mutant animals pituitary GnRH receptor levels are approximately 30% less than in normal mice, their number can be rapidly increased by GnRH injections (Young *et al.*, 1982), so that the pituitary gland of hpg mutant mice responds normally to GnRH administration with LH release. LH and FSH receptors are also present in the gonads of the hpg mutant, and it has been shown that in the hpg mutant male, when LH and/or FSH are administered as multiple doses over a 24-hour period, the testes respond with enhanced androgen production and spermatogenesis. The gonads of both sexes develop normally when transplanted to normal animals (Bamber *et al.*, 1980). Preliminary studies of feedback responsiveness in hpg mutant animals indicate that the hypothalamus is a major site of testosterone negative feedback, while that of estradiol seems to be (in contrast to normal animals) at the pituitary level.

It is not known whether there are alterations on the content of other neuropeptides or neurotransmitters in such hpg mutant animals.

Factors Involved in Functional Transplants of CNS Tissue

Success of such transplantation has been ascribed to the presence of the blood-brain barrier, which presumably, by keeping cells of the immune system from reaching the brain, makes the brain an immunologically privileged site. Available data indicate that early fetal tissue works best, probably related to the ability of such donor tissue to continue to mature and divide at its own intrinsic rate (rather than at that of the host) (Das, 1980; Jaeger and Lund, 1980a; Jaeger and Lund, 1980b; Lund and Harvey, 1981; McLoon *et al.*, 1981; Stenevi *et al.*, 1980b). Critical studies with regard to host age are still required; many of the above studies have been performed in young adult as well as fully adult hosts with apparently equal success. There is also information (in abstract form) that transplantation is successful across strains (Low *et al.*, 1982) and across species (Stenevi *et al.*, 1982) (see also Kolata, 1982).

Axonal relationships between the host and a variety of transplants have been anatomically and to some extent physiologically defined, with considerable variability as to what has been observed to date. Host afferents can grow accurately to specific areas within the transplant even though they are not normally afferent to the transplanted tissue (Dunnett, 1981; Jaeger and Lund, 1980a; Oblinger *et al.*, 1980; Perlow *et al.*, 1979). Likewise, when transplanted adjacent to the normal terminal region, transplant efferents can project to the host in a normal (Bjorklund *et al.*, 1976; McLoon and Lund, 1980) or an abnormal manner (Bjorklund *et al.*, 1976). When placed into a heterotopic location such efferents can be very abnormal, reflecting neither the transplanted tissue nor the host region (Jaeger and Lund, 1980a; Oblinger *et al.*, 1980). A major question is to what extent host-transplant axonal relationships are influenced by tissue-specific factors, serving as physical guidance systems and/or biochemical factors with trophic activity (Fraser and Hunt, 1980; Katz and Lasek, 1980; Nieto-Sampedro *et al.*, 1982; Varon and Bunge, 1978).

The studies to be described are still preliminary and have to date considered only a small number of the questions that can be addressed with the experimental model employed--based on the considerations cited above--that of transplantation of fetal preoptic area (POA) (containing GnRH cell bodies) into GnRH-deficient hpg mutant mice. The questions to be considered are:

1. Can the hypogonadism of male and female hpg mutant mice be corrected by transplants of normal fetal POA (from unaffected littermates) into the brains of the hpg aminals?
2. If such transplants are successful, will (a) such transplants be effective across strains and species, and (b) will POA transplants from older donors (i.e., neonatal rather than fetal) be effective in correcting the GnRH deficiency?
3. Is the sex of the donor animal, compared to that of the recipient, a significant factor in the success of a transplant in correcting the GnRH deficiency?
4. What is the nature of the afferent and efferent projections to and from successful implants?

METHODS AND RESULTS

Effect of Fetal POA Transplants from Unaffected Hpg Animals in Correcting the Hypogonadism of the Adult Male *Hpg Mouse*

These experiments studied the effect of implantation of fetal (16-18 day old fetuses) preoptic area (a site of GnRH production) from unaffected hpg animals on hypothalamic-pituitary-gonadal function in adult mutant (5-6 month old male) hpg mice. For control tissue transplants to hpg and normal animals, cortex from prepubertal mice was utilized. Two preoptic (POA) tissue segments were used per implant; cortical fragments were of the same dimensions as the preoptic fragments. Transplants in 2-4 μl saline were stereotaxically placed via a 22-gauge needle into the anterior third ventricle of anesthetized recipients. There were five experimental groups: normal animals, hpg mutant animals, hpg mutant animals with POA transplants, hpg mutant animals with transplants of cortex, normal animals with transplants of cortex.

The growth rates of grafted animals were similar to those of unaffected mutants or normal animals. Two months post implantation blood was obtained via retro-ocular puncture for plasma gonadotropin and testosterone determinations, and the animals then killed by decapitation. On inspection at time of sacrifice, it was apparent that of the hpg mutant animal groups, only in those with POA implants were testes descended into the scrotum. In a given experimental group, half of the brains were processed for immunocytochemical studies, the other half for GnRH immunoassay (Nett *et al.*, 1973). Immunoassays were performed on sections which consisted of hypothalamus alone, and on another pool which represented remaining brain sections, excluding the brainstem from the midbrain down. For immunocytochemical studies, brains were divided into three coronal blocks, the first coronal cut being made a millimeter in front of the optic chiasm and the second behind the mammillary bodies. The anti-GnRH serum employed has previously been reported (Schwanzel-Fukuda and Silverman, 1980).

Pituitary glands were bisected *in situ*, one half of each being frozen until time of assay for gonadotropin concentrations, the other processed for future immunocytochemical studies.

Figure 1 indicates the effect of preoptic area or control cortical implants on hypothalamic GnRH concentrations, testicular weight, and pituitary and serum gonadotropin concentrations. Evidence of functional grafts was found in 7 of the 8 POA-hpg mutant implanted animals, while none was present in hpg mutant animals implanted with cortex. It is apparent from Fig. 1 that such grafts do not totally correct GnRH, LH, or FSH (pituitary and plasma) levels to those seen in normal animals. [Whether this can be achieved by implantation of larger amounts of preoptic area tissue remains to be seen. It is also possible (see studies of female animals) that allowing a longer time period for the graft to grow within the host would make for greater correction of the hormonal abnormalities present in the hpg mutant. Another possibility is that lack of complete normalization is secondary to the absence of appropriate afferent connections to transplanted GnRH cells.] Plasma testosterone concentrations were also incompletely corrected [(ng ml^{-1}): normal 1.0, hpg and hpg + cortex both undetectable, and hpg + POA 0.3]. These testosterone levels correlated with the observed differences in seminal vesicle weight, used as an index of biological testosterone efficacy. Such weights (mg per pair) were: normal, 258.3 ± 25.6; normal + cortex, 218.0 ± 24.6; hpg, 17.2 ± 1.6; hpg + cortex, 14.0 ± 2.9; and hpg + POA, 66.1 ± 16.2.

Testicular weight (Fig. 1B, Fig. 2) was also corrected to an intermediate extent in POA + hpg mutant implanted animals. Histological examination of the testes revealed full spermatogenesis in the hpg mutant + POA recipients, while hpg mutant animals with cortical implants were indistinguishable from those of untreated hpg animals (Fig. 3).

Studies of brain sections correlated well with individual observations of grafted animals. In the single animal without hormonal evidence of a functional transplant, a cannula scar, but no evidence of transplanted tissue, was seen. In all four remaining POA-grafted hpg mutant animals that were examined immunocytochemically, evidence of viable grafts containing GnRH immunoreactive neurons was present (Fig. 4).

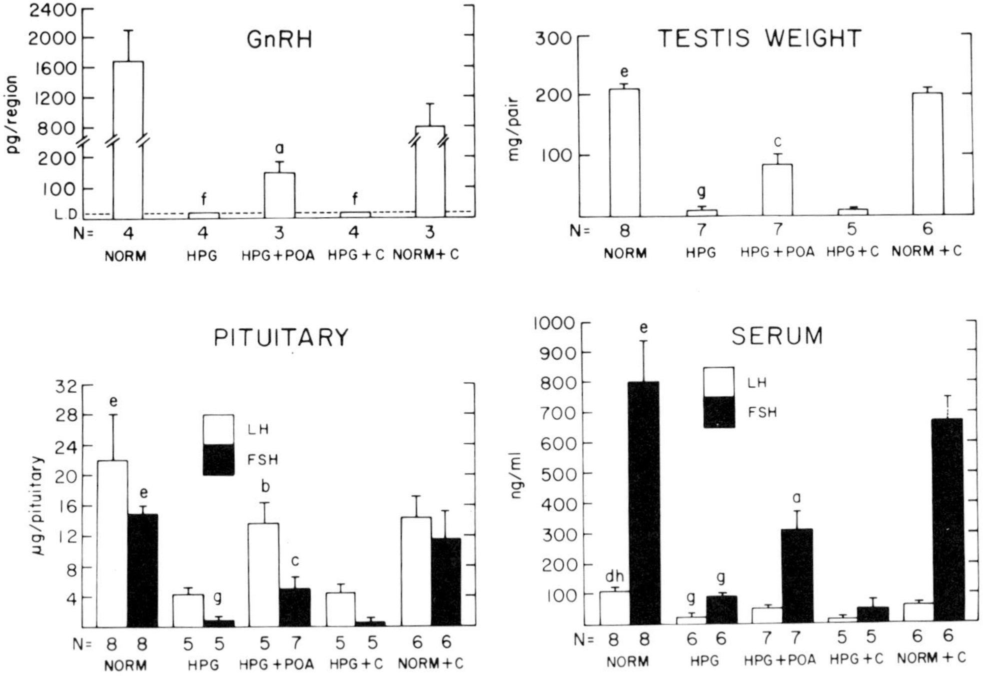

FIGURE 1. Effect of implants of preoptic area (POA) or cortex (C) on hypothalamic GnRH concentrations (A), testicular weight (B), and pituitary (C) and serum (D) gonadotropin concentrations. HPG, mutant hypogonadal mouse; NORM, normal littermate. Bars indicate mean ± SEM. LD, limits of detection. GnRH values for hpg and hpg + cortex were combined, as they were the same. POA + hpg compared with hpg and hpg + C: a = P<0.05; b = P<0.02; c = P<0.01. NORM compared with hpg + POA: d = P<0.05; e = P<0.001. NORM compared with hpg: f = P<0.05; g = P<0.001. NORM compared with NORM + C: h = P<0.05.
Reproduced from Krieger et al., 1982, with permission.

FIGURE 2. Effects of implants of preoptic area on testicular size. Testis on left is from an adult hpg mutant mouse, that on right from a normal littermate. In the middle is a testis from an hpg mutant animal that received a fetal preoptic area implant.

Perikarya immunostained for GnRH were present in all of the grafts; immunoreactive fibers could be seen entering the median eminence in its lateral aspects, with some projecting to capillaries in the zona externa. No evidence of GnRH staining was found in untransplanted hpg animals or those with cortical transplants. The normal adult mice similarly showed immunoreactive GnRH fibers in the median eminence, particularly in its lateral aspects--i.e., a location similar to the site of projection of fibers from the POA grafts. No GnRH-reactive perikarya were present in any of the cortical transplants. Outgrowths from cortical grafts infiltrated the surrounding area but did not grow to the median eminence.

Gonadotropin receptor concentrations were also determined in 20,000 g fractions of decapsulated testicular homogenates obtained from normal animals, hpg mutants, and hpg with preoptic area implants, using ^{125}I-hCG and receptor-purified ^{125}I-hFSH in Dulbecco's phosphate buffered saline. Homogenates were incubated overnight at room temperature under conditions approximating saturation and the supernatants assessed individually for testosterone by radioimmunoassay. The results indicated that there were marked reductions in the number of testicular hCG and FSH receptors in affected hpg mice, which increased (though not to normal levels) in animals with functional POA grafts.

These data indicate that partial restoration of hypothalamic GnRH levels in hpg mutant animals (deficient in such GnRH) can be produced by grafted GnRH cells. Such restoration brings about release of sufficient pituitary LH and FSH to

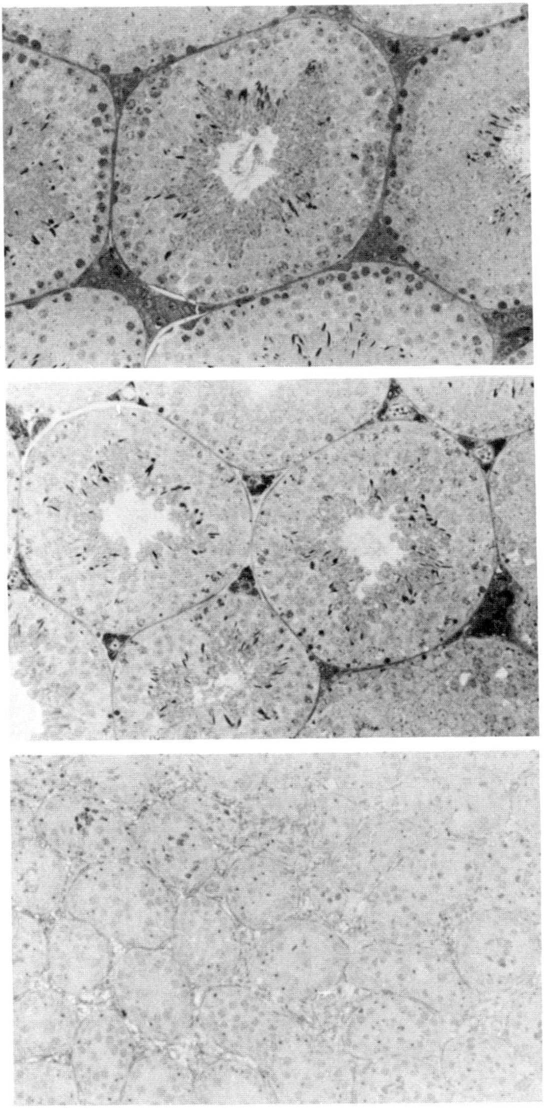

FIGURE 3. Testicular sections from normal animal (top); hpg animal with POA implant (middle); and hpg animal with cortex implant (bottom). Full spermatogenesis and evidence of interstitial cell development is present in POA recipients, while sections from animals with cortical implants are identical to those from untreated hpg animals. x 154. Reproduced from Krieger et al., *1982, with permission.*

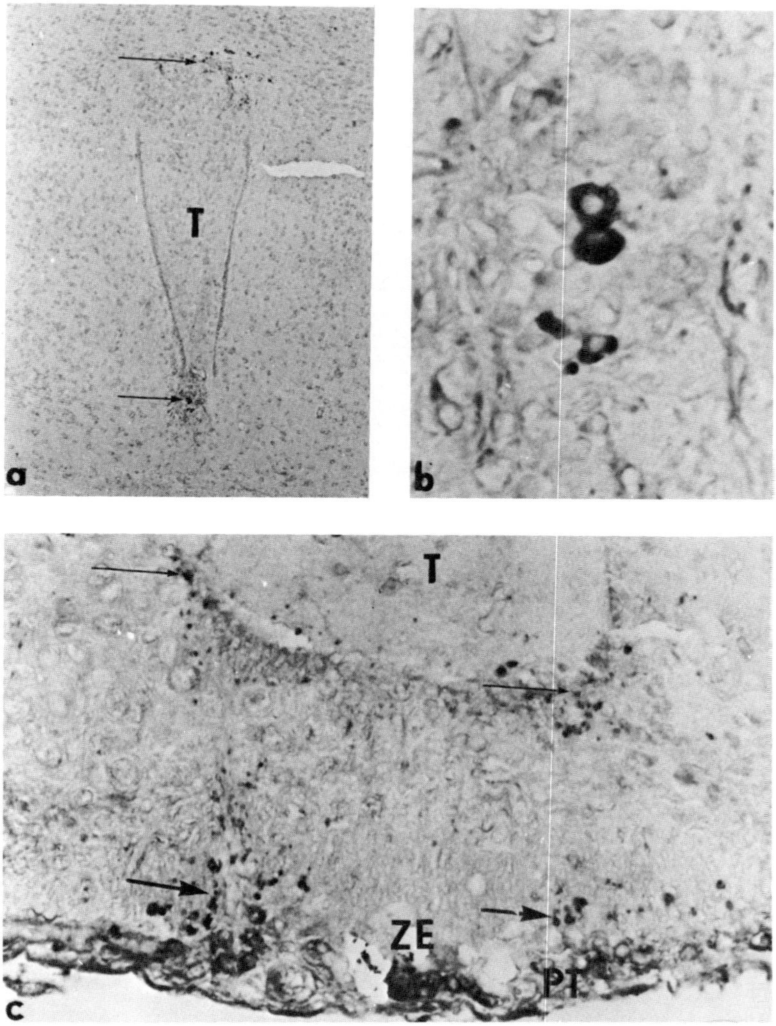

FIGURE 4. Localization of GnRH in a hpg mouse hypothalamus containing a third ventricular transplant (T) of normal fetal mouse preoptic area. a, coronal section at the level of the mid-hypothalamus just rostral to the median eminence. The transplant fills the third ventricle and extends into the brain tissue just dorsal and ventral of it (arrows), where reactive neuronal perikarya were found. b, higher magnification of three reactive neurons indicated by the lower arrow in a. c, section more caudal to a, again showing the transplant within the third ventricle at the level of the mid-

cause greatly increased testicular weights and normal spermatogenesis and interstitial secretory activity.

The data also indicate, therefore, the presence of intact and functioning pituitary GnRH receptors and also receptivity of the previously undeveloped testicular tissue, as evidenced by development of gonadotropin receptors therein (Davies *et al.*, 1982).

Effect of Fetal POA Transplants from Unaffected Hpg Animals in Correcting the Hypogonadism of the Adult Female *Hpg Mutant Mouse*

These studies were performed in a manner similar to those described for the male animals. Successful transplants have been achieved in 60% of implanted mutant hpg female animals, as gauged by vaginal opening, persistently cornified vaginal smears, increases in reproductive tissue weight, and increased plasma and pituitary gonadotropin concentrations. In animals with successful implants, the longer the implant was present (i.e., 4 months compared to 1 month), the more closely did the endocrine parameters observed resemble those of normal hpg animals. Although the ovaries showed evidence of marked follicular development in animals with successful implants, none contained corpora lutea. Two of the successfully implanted female animals, when caged with normal males, mated without becoming pregnant.

None of the hpg mutant animals bearing implants of cortical tissue manifested any correction of their hypogonadal state.

Effect of Strain and Species of Donor on Success of Implantation

Limited studies have been performed in both male (implantation of fetal POA from Swiss-Webster mice) and female hpg mutants (implantation of fetal POA from rat or vole). These studies thus far indicate that implants may be successful across strains but not across species.

---

median eminence. Reactive fibers enter the dorsal median eminence at the ventrolateral corners of the graft (upper smaller arrows), and some can be traced to the zona externa (ZE) (lower, larger arrows). Some background staining can be seen in the pars tuberalis (PT) just external to the ZE. a, x 112; b, x 764; c, x 360. Reproduced from Krieger et al., *1982, with permission.*

Effect of Age of Donor on Success of Implantation

Studies to date in which donor tissue derived from 15-day-old normal hpg animals was implanted into male or female hpg mutant animals revealed no evidence of pituitary or gonadal stimulation.

Is the Sex of the Donor Animal Compared to That of the Recipient a Significant Factor in the Success of a Transplant in Correcting the GnRH Deficiency in the Hpg Mutant?

These studies were undertaken in view of the known sexual dimorphism of the POA. The initial report of Raisman and Field (1971) indicating sexual dimorphism in the ratio of synapses on dendritic shafts to the number of dendritic spines in the POA led the way to a larger literature confirming such dimorphism with regard to other aspects of CNS organization, i.e., POA nuclear volume, neuronal dendritic branching, and synaptic organization (see MacLusky and Naftolin, 1981, for review). It is now established that the mammalian brain is inherently female or bipotential, with sexual differentiation of its reproductive functions largely determined by testicular hormones secreted during perinatal development in the male. It is felt that aromatization of testosterone to estrogen through an interaction with estrogen receptors in the medial preoptic area is critical in this regard (Naftolin *et al.*, 1975; Plapinger and McEwen, 1978). Such "masculinization" of the brain results in permanent suppression of the capacity to support cyclic feminine patterns of gonadotropin secretion and suppression of a number of other behavioral and neural endocrine patterns characteristic of the female, with enhancement of patterns characteristic of the male. The anatomic sex differences noted may represent expression of the growth-promoting effects of gonadal steroids analogous to those observed in culture (Toran-Allerand, 1976). The major period of cytoplasmic differentiation and synaptic formation in the rat POA begins by the time of birth and continues throughout the first days of postnatal life (MacLusky and Naftolin, 1981), after the critical period for sexual differentiation.

In view of these observations, the question arose as to whether the lack of cyclicity in female hpg mutant animals with functioning transplants was secondary to the implantation of a "masculine" POA, even though the age of the fetal donors employed may have been slightly younger than that noted above for time of appearance of the major dimorphic changes (see Discussion).

What Is the Nature of the Afferent and Efferent Projections To and From Successful Implants?

With regard to efferent projections from transplanted GnRH cells, it has already been noted that these projected to the same lateral area of the median eminence in which GnRH fibers from normal animals were observed to terminate. To date, such studies have been performed only on 6 μm Bouin's immersion-fixed paraffin sections. In these studies, no GnRH projections from the graft to extrahypothalmic areas were noted; if such were present, they might not be visible secondary to the tissue losses of antigen known to be associated with organic solvents used with paraffin, or the limited depth of field available with such thin sections, which compromises the ability to trace fiber tracts. Additional studies utilizing thicker Vibratome sections will be necessary to fully elucidate details of GnRH fiber distribution.

With regard to afferent input to POA implants, limited studies have been performed assessing adrenergic innervation, as reflected by immunostaining with a tyrosine hydroxylase antibody kindly provided by Dr. T. Joh. These studies have also been performed, to date, only on the Bouin's immersion-fixed paraffin sections. They indicate both the presence of positive cell bodies within the graft as well as positive fibers which appear to enter the graft from the same portion of the median eminence to which the GnRH fibers exit.

DISCUSSION

The present studies provide an affirmative answer to the first question asked, namely, that transplantation of fetal preoptic area--the location of GnRH-positive neurons--into hypogonadal male and female mutant animals known to lack hypothalamic GnRH can remedy this neurochemical defect with regard to correction of a number of abnormal hormonal parameters present in the mutant animals. In both male and female GnRH-deficient hpg mice that received POA transplants which were shown to be functioning, there was evidence of increased hypothalamic GnRH content (as compared to its absence in the untreated mutant animals), and outgrowth of GnRH fibers from implants to the median eminence was demonstrated. There were increases in pituitary LH and FSH content and plasma LH and FSH concentrations in mutant animals of both sexes with POA transplants. In the male mutant animals, there was elevation of plasma testosterone concentrations and evidence of testicular growth and scrotal descent, accompanied by histological evidence of full spermatogenesis and of interstitial cell development. In female animals, there was marked increase in ovarian and uterine size, evidence of vaginal opening and estrous smears, and mating behavior. It should be stressed that in the male mutant animals the POA transplants did not completely restore reproductive tissue weight or pituitary plasma gonadotropins and plasma testosterone levels to normal levels. As already noted, this may be related to the insufficient time in which the transplant was present (i.e., 2 months), in contrast to the more effective normalization of ovarian and uterine weight and pituitary gonadotropin concentrations in female mutant animals, in whom POA transplants were present for a longer time period. It is also possible that factors such as the amount of material transplanted and the efficacy of innervation of the transplants (see below) are other factors to be considered with regard to the incompleteness of response.

The lack of cyclic vaginal smears, the absence of corpora lutea, and the preliminary evidence of inability to conceive in the female mutant animals in whom POA transplants had achieved evidence of increased pituitary-gonadal function, raises the question as to whether or not appropriate innervation which may be necessary for cyclic hormone release, is present in such transplants. The preliminary evidence we have presented of catecholaminergic innervation of the transplants requires further confirmation at the EM level, in addition to the need to delineate possible contributions by other neurotransmitter agents, which may differ in the grafted and in the intact normal hpg animals. The data also provide preliminary information with regard to some of the other questions asked. It appears that tissue from older animals (i.e. preoptic area derived from 15-day-old animals) is not effective in correcting the hormonal abnormalities in the hypogonadal hpg mutant animal; additional studies are required at earlier neonatal ages. It also appears that fetal POA transplants derived from other strains of mice, but not from other species, as as effective as syngeneic fetal transplants. To date, there is no evidence that the sex of the animal from which the fetal transplant was derived is an important factor with regard to the efficacy of such transplants. Should early neonatal transplantation be effective, this may have to be reconsidered, in view of the greater sexual dimorphic commitment in such older animals.

The observation that there appears to be tissue site-targeted outgrowth of GnRH fibers from the implanted POA to the lateral aspect of the median eminence raises the question of a possible role of trophic factors in the host tissue which implement the direction of graft outgrowth. It is now well recognized that there are both region- and cell-specific associations between various CNS areas and their target tissues (Attardi and Sperry, 1963; Edds, 1967; Lund, 1978). The mechanisms by which axons go to one region rather than to another are poorly understood. This seems to be dependent on information in both the pre- and postsynaptic cells. A number of specific neuronal growth factors have been characterized (Levi Montalcini, 1975; Nieto-Sampedro *et al.*, 1982). There is also evidence that surface molecules are involved in the interactions among neural cells during development (Rutishauser *et al.*, 1979). Whether such factors are involved in influencing POA outgrowth in the hpg mutant animal remains to be determined.

The present findings of the prolonged maintenance of relatively normal hormone parameters and the high success rate achieved in the transplanted animals in this study indicate that the POA-transplanted hpg mutant animal can answer fundamental questions with regard to the efficacy of the brain transplant, possible role of sexually dimorphic regions, factors involved in pulsatile and cyclic gonadotropin secretion, and the role of trophic factors in determining the sites of nerve cell outgrowth.

REFERENCES

1. Attardi, D.G., and Sperry, R.W., *Exp. Neurol. 7,* 46, (1963).
2. Azmitia, E.C., Perlow, M.J., Brennan, M.J., and Lauder, M.M., *Brain Res. Bull. 7,* 703 (1981).
3. Bamber, S., Iddon, C.A., Charlton, H.M., and Ward, B.J., *J. Reprod. Fertil. 58,* 249 (1980).
4. Bjorklund, A., and Stenevi, U., *Brain Res. 229,* 403 (1981).
5. Bjorklund, A., Stenevi, U., Dunnett, S.B., and Iversen, S.D., *Nature 289,* 497 (1981).
6. Bjorklund, A., Dunnett, S.B., Stenevi, U., Lewis, M.E., and Iverson, S.D., *Brain Res. 199,* 307 (1980).
7. Bjorklund, A., and Stenevi, U., *Physiol. Rev. 59,* 62 (1979).
8. Bjorklund, A., Kromer, L.F., and Stenevi, U., *Brain Res. 173,* 57 (1979).
9. Bjorklund, A., Stenevi, U., and Svendgaard, N.-A., *Nature 262,* 787 (1976).
10. Carmel, P.W., Araki, S., and Ferin, M., *Endocrinology 99,* 243 (1976).
11. Das, G.D., *Life Sci. 4,* 93 (1974).

12. Das, G.D., Hallas, B.H., and Das, K.G., *Am. J. Anat. 158,* 135 (1980).
13. Davies, T.F., Platzer, M., Perlow, M.J., Gibson, M., Zimmerman, E.A., Ferin, M., and Charlton, H.M., *Proc. 64th Ann. Meeting Endocrine Soc.,* p. 320, #963 (1982).
14. Dunnett, S.B., Bjorklund, A., Stenevi, U., and Iversen, S.D., *Brain Res. 215,* 147 (1981).
15. Edds, M.V., Jr., *in* The Neurosciences: A Study Program, p. 230, Rockefeller U. Press, New York (1967).
16. Eskay, R.L., Mical, R.S., and Porter, J.C., *Endocrinology 100,* 263 (1977).
17. Foremann, M.M., and Moss, R.L., *Horm. Behav. 8,* 219 (1977).
18. Fraser, S.E., and Hunt, R.K., *Ann. Rev. Neurosci., 3, 319* (1980).
19. Freed, W.J., Perlow, M.J., Karoum, F., Seiger, A., Olson, L., Hoffer, B.J., and Wyatt, R.J., *Ann. Neurol. 8,* 510 (1979).
20. Gash, D., Sladek, J.R., and Sladek, D.C., *Science 210,* 1367 (1980).
21. Gibson, M.J., Perlow, M.J., Zimmerman, E.A., Davies, T.F., Charlton, H.M., Ferin, M., and Krieger, D.T., *Proc. 65th Ann. Meeting Endocrine Soc.* (in press) (1983).
22. Hallas, B.H., Das, G.D., and Das, K.G., *Am. J. Anat. 158,* 147 (1980).
23. Jaeger, C.B., and Lund, R.D., *J. Comp. Neurol. 194,* 571, 1980a.
24. Jaeger, C.B., and Lund, R.D., *Exp. Brain Res. 40,* 265, 1980b.
25. Jennes, L., and Stumpf, W.E., *Neuroendocrinol. Lett. 2,* 241 (1980).
26. Kalra, S.P., Simpkins, J.W., and Kalra, P.S., *Acta Endocrinol. (Copenh.) 95,* 1 (1980).
27. Katz, M.J., and Lasek, R.J., *Cell Motil. 1,* 141 (1980).
28. Kolata, G., *Science 217,* 342 (1982).
29. Krieger, D.T., Perlow, M.J., Gibson, M.J., Davies, T.F., Zimmerman, E.A., Ferin, M., and Charlton, H.M., *Nature 298,* 468 (1982).
30. Kromer L.F., Bjorklund, A., and Stenevi, U., *Brain Res. 210*, 153 (1981a).
31. Kromer, L.F., Bjorklund, A., and Stenevi, U., *Exp. Brain Res. 210,* 173 (1981b).
32. Kromer, L.F., Bjorklund, A., and Stenevi, U., *Science 204,* 1117 (1979).
33. Levi-Montalcini, R., *in* The Neurosciences; Paths of Discovery, p. 245, M.I.T. Press, Cambridge (1975).

34. Low, W.C., Lewis, P.R., and Bunch, S.T., *Proc. Soc. for Neurosci. 8,* 747 (1982).
35. Low, W.C., Dunnett, S.B., Bunch, S.T., Thomas, S.R., Lewis, P.R., Iversen, S.D., Bjorklund, A., and Stenevi, U., *Neurosci. Abst. 7,* 259 (1981).
36. Lund, R.D., Development and Plasticity of the Brain. Oxford U. Press, New York (1978).
37. Lund, R.D., and Harvey, A.R., *J. Comp. Neurol., 201,* 191 (1981).
38. Lund, R.D., and Hauschka, S.D., *Science 193,* 582 (1976).
39. MacLusky, N.J., and Naftolin, F., *Science 211,* 1294 (1981).
40. McLoon, S.C., and Lund, R.D., *Exp. Brain Res. 40,* 273 (1980).
41. McLoon, L.K., McLoon, S.C., and Lund, R.D., *Brain Res. 226,* 15 (1981).
42. Merchenthaler, I., Kovacs, G., Lovasz, G., and Setalo, G., *Brain Res. 198,* 63 (1980).
43. Naftolin, F., Ryan, D.J., Davis, I.J., Reddy, V.V., Flores, F., Petro, Z., Kuhn, M., White, R.J., Takaoka, Y., and Wolin, L., *Rec. Prog. Horm. Res. 31,* 295 (1975).
44. Nett, T.M., Akbar, A.M., Niswender, G.D., Hedlund, M.T., and White, W.F., *J. Clin. Endocrinol. Metab. 36,* 880 (1973).
45. Nieto-Sampedro, M., Lewis, E.R., Cotman, C.W., Manthrope, M., Skaper, S.C., Barbin, G., Longo, F.M., and Varon, S., *Science 214,* 860 (1982).
46. Oblinger, M.M., Hallas, B.H., and Das, G.D., *Brain Res. 189,* 228 (1980).
47. Olson, L., Seiger, A., Taylor, D., Freedman, E., and Hoffer, B.J., *Exp. Brain Res. 39,* 277 (1980a).
48. Olson, L., Bjorklund, H., Ebendal, T., Hedlund, K.O., and Hoffer, B., *Ciba Foundation Symposium 83,* 213 (1980b).
49. Palkovits, M., *Neuroendocrinology 33,* 123 (1981).
50. Perlow, M.J., Freed, W.J., Hoffer, B.J., Seiger, A., Olson, L., and Wyatt, R.J., *Science 204,* 643 (1979).
51. Pfaff, D.W., *Science 182,* 1148 (1973).
52. Plapinger, L., and McEwen, B.S., *in* Biological Determinants of Sexual Behavior, p. 153, John Wiley, New York, (1978).
53. Raisman, G., and Field, P.M., *Science 173,* 731 (1971).
54. Rance, N., Wise, P.M., Selmanoff, M.K., and Barraclough, C.A., *Endocrinology 108,* 1975 (1981).
55. Ratner, A., *Proc. Soc. Exp. Biol. Med. 138,* 995 (1971).
56. Rutishauser, U., Brackenbury, R., Thiery, J.-P., and Edelman, G.M., *Birth Defects 15,* 79 (1979).

57. Sakuma, Y., and Pfaff, D.W., *Nature 283,* 566 (1980).
58. Sawyer, C.H., Markee, J.E., and Hollinshead, W.H., *Endocrinology 41,* 395 (1947).
59. Scholer, J., and Sladek, J.R., Jr., *Science 214,* 347 (1981).
60. Schwanzel-Fukuda, M., and Silverman, A.J., *J. Comp. Neurol. 191,* 213 (1980).
61. Segal, M., Stenevi, U., and Bjorklund, A., *Neurosci. Lett. 27,* 7 (1981).
62. Stenevi, U., Bjorklund, A., Dunnett, S.B., and Gage, F.H., *Proc. Soc. for Neurosci. 8,* 748 (1982).
63. Stenevi, U., Bjorklund, A., Kromer, L.F., Paden, C.M., Gerlach, J.L., McEwen, B.S., and Silverman, A.J., *Cell Tiss. Res. 205,* 217 (1980a).
64. Stenevi, U., Bjorklund, A., Kromer, L.F., Paden, C.M., Gerlach, J.L., and McEwen, B.S., *Cell Tiss. Res. 205,* 217 (1980b).
65. Toran-Allerand, C.D., *Brain Res. 106,* 407 (1976).
66. Turgeon, J., and Barraclough, C.A., *Endocrinology 92,* 755 (1973).
67. Varon, S.S., and Bunge, R.P., *Ann. Rev. Neurosci. 1,* 327 (1978).
68. Wells, J., and McAllister, J.P., II, *Brain Res. 4,* 167 (1982).
69. Witkins, J.W., Paden, C.M., and Silverman, A.J., *Neuroendocrinology 35,* 429 (1982).
70. Young, L.S., Charlton, H.M., and Clayton, R.M., *Proc. 1st Joint Meeting British Endocrine Societies,* p. 58, #116 (1982).

PART IV

Posttranslational Processing

POSTTRANSLATIONAL PROCESSING OF GASTRIN/CCK PEPTIDES IN THE CENTRAL NERVOUS SYSTEM

Jens F. Rehfeld

University Department of Clinical Chemistry
Rigshospitalet, Copenhagen, Denmark

The gut hormones gastrin and cholecystokinin (CCK) with the same C-terminal active site are both synthesized in the brain. While gastrin occurs only in hypothalamo-hypophyseal neurons, CCK is distributed to all regions but epithalamus and cerebellum. Human gastrin is processed from a 101 AA precursor to component I (~ 45 AA), gastrin-34 and gastrin-17, which all occur in sulphated and non-sulphated forms. While this processing resembles that of gastric G-cells, it differs from that of corticotrophs, where only non-sulphated component I and gastrin-34 occur. We have not yet data on CCK-mRNA. However, time- and pulse-chasing with ^{35}S-methionine in vivo in rats reveal a rapid synthesis of CCK in the cerebral cortex. Moreover, we have now identified eight major molecular forms of sulphated CCK from mammalian brains. In most regions CCK-8 predominates. The many brain CCKs suggest either unusually many steps in posttranslational processing in each neuron or a differential processing in different CCK-neurons.

INTRODUCTION

Gastrin and cholecystokinin (CCK) are homologous gut hormones discovered early in this century. They share the C-terminus -Trp-Met-Asp-Phe-NH_2, which constitutes their active site. The site is preserved during the evolution from coelenterates (Grimmelikhuijzen et al., 1980) to man, and apparently gastrin and CCK have developed from a common ancestor (Larsson and Rehfeld, 1977). The identical active site gives a functional overlap, in which the potency of each hormone is determined by the

BIOCHEMICAL AND CLINICAL ASPECTS OF NEUROPEPTIDES: SYNTHESIS, PROCESSING, AND GENE STRUCTURE

ISBN 0-12-417320-9

different N-terminal extensions. Hence, functional, structural and evolutionary features argue that gastrin and CCK are one regulatory system.

In 1975 Vanderhaeghen et al. discovered peptide-like material in the CNS reacting with C-terminal specific gastrin antibodies. Subsequent studies showed the material in most regions to be CCK (Dockray, 1976; Rehfeld, 1977; 1978a). However, hypophyseal neurons contained true gastrin (Rehfeld, 1978a). By now octapeptide CCK has been shown to fulfill most neurotransmitter criteria (Rehfeld et al., 1979). Moreover, CCK in contrast to most other neuropeptides occur in nearly all CNS regions with the highest concentrations in cerebral cortex (Rehfeld, 1978). Thus, CCK is so far the most abundant active neuropeptide in the human brain, while gastrin as mentioned occurs only in discrete hypothalamo-hypophyseal neurons (Larsson and Rehfeld, 1981). The exact roles and clinical significance of neuronal gastrin and CCK are still unknown. However, their well-defined occurrence and considerable transmitter potential justify the continued exploration of their biology.

The following review describes the present knowledge about the posttranslational processing of cerebral gastrin and CCK as obtained in our laboratories.

METHODS

Three different strategies have been used to obtain information about the posttranslational processing.

I. Molecular Identification of Different Gastrin/CCK-Peptides in Different Regions of The Brain

After dissection three principles have been employed. First, *differential extraction techniques* have proven necessary to yield maximal removal from the tissue with minimal modifications of the peptides (Rehfeld, 1978b; Marley and Rehfeld, 1983). In order to obtain maximal release of the known gastrins and of the largest molecular form of CCK, CCK-8 and smaller C-terminal fragments, boiling in water of frozen tissue pieces (weighing a few mg) has proven optimal. Reextraction in cold 0.5 M acetic acid has subsequently yielded CCK-forms of intermediate size (from ~ 60 to 12 amino acids). Cold 90% methanol: 10% H_2O has proven ideal for selective extraction of sulphated CCK-8 from brain tissue. Other allegedly used extraction conditions (HCl, NaOH or C_2H_5OH) have proven either disastrous to the peptide structure or insufficient (Marley and Rehfeld,

1983).

After the differential extraction, the molecular identification has been further approached by *chromatography*: We have used gel filtration on columns of a variety of dimensions and gel qualities, subsequent cellulose ion exchange chromatography and recently reverse-phase or ion exchange HPLC of components isolated by gel or cellulose chromatography. For selected problems we have in addition used immunoaffinity chromatography, re-chromatography in urea gradients or dithiothreitol and rechromatography after treatment with trypsin, clostripain, carboxypeptidase B, sulfatases and other enzymes.

At the analytical level, extractions and chromatographies have been monitored by *radioimmunoassays*. Due to the structural homology between gastrin and CCK-peptides, and to the pronounced heterogeneity comprising a multitude of peptides of different chain lengths as well as amino acid derivatizations (sulphation of tyrosyl, C-terminal amidation of phenylalanyl and N-terminal cyclization of glutamyl to pyroglutamyl) we have developed two techniques above the conventional level of radioimmunochemical specificity: Sequence-specific radioimmunoassays, which is a panel of assays specific for *different* sequences of the peptides (Rehfeld, 1978b), and more recently residue-specific radioimmunoassays (Rehfeld and Morley, 1983). The latter is an entirely novel analytical tool, which consists of a panel of radioimmunoassays directed against the *same* sequence of 4-5 amino acid residues, but where the antibodies display different reactivities for each residue. Hence, it is possible to detect modifications of a single amino acid residue.

II. In Vivo Time- and Pulse-Chase Studies of The Biosynthesis of Cerebral CCK-Peptides

In rats labelled amino acids, mainly ^{35}S-methionine, were injected intracerebroventricularly in large doses. The rats were decapitated at intervals from 15 to 600 min with and without pulse-chasing (Goltermann et al., 1980a and b). The brains were rapidly removed, dissected on a cold plate and extracted as described above. Using a C-terminal specific immunosorbens all the biological active CCKs (i.e. containing the C-terminal active tetrapeptide amide) were removed from the extracts, displaced from immunosorbens by unlabelled gastrin-17 and applied to gel chromatography monitored by ^{35}S, ^{14}C or ^{3}H counting as well as radioimmunoassays specific for the nonhomologous sequences of CCK.

III. Isolation of mRNA for Studies of Gene and Peptide Precursor Structure

Using the oligodeoxynucleotide probe technique of Noyes et al. (1979) we have recently isolated the gastrin mRNA from a human gastrinoma (Boel et al., 1983). The subsequently constructed cDNA has been cloned, and used to determine the coding region (303 nucleotides) of the human gastrin gene. From this region the sequence of the primary translation product has been determined (Boel et al., 1983). The precursor is a polypeptide of 101 amino acids. Since examination of a human DNA-library so far has revealed only one gastrin gene, we conclude that the neuronal gastrins are processed from the same precursor as that of all other gastrin producing cells. In spite of great efforts, our attempts to isolate CCK-mRNA from brain or gut tissue have so far been unsuccessful. However, we still hope to succeed.

RESULTS AND DISCUSSION

Posttranslational Processing of Gastrin

The gastrin precursor, component I, gastrin-34 and gastrin-17 occur in discrete hypothalamo-hypophyseal (sulphated as well as non-sulphated) neurons (Rehfeld, 1978a; Larsson and Rehfeld, 1981; Rehfeld and Larsson, 1981). Due to concentrations and molecular pattern in different species, uptake from blood cannot explain the occurrence. Moreover, the concentrations (~ 20 pmol/g) and molecular pattern vary - in contrast to CCK - very little in mammals indicating that hypothalamo-hypophyseal gastrinergic neurons may be of functional significance (Rehfeld et al., submitted 1983). The gastrin immunoreactivity originates apparently in some magnocellular neurons of the supraoptic and paraventricular hypothalamic nuclei (Vanderhaeghen et al., 1981; Larsson et al., unpublished). If we combine the observation, that only one gastrin gene coding for one 101 AA precursor occurs in humans with the observed molecular pattern of hypothalamo-pituitary gastrins, the posttranslational processing of biological active gastrins in the central nervous system follows the scheme below (asterisk indicates the position of sulphated tyrosyl):

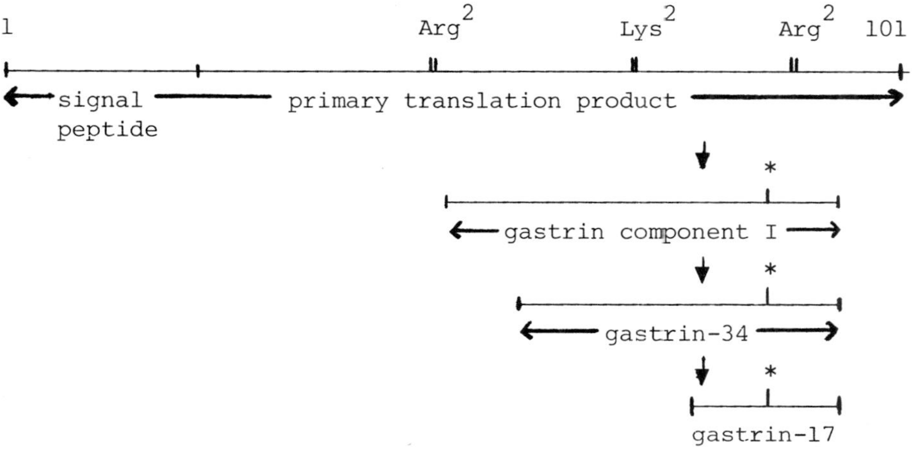

The processing corresponds to that of the G-cells in the stomach, which contain the same precursor products. In both cases approximately 40% of each gastrin is sulphated. However, we still do not know whether neuronal gastrin-17 is processed further to gastrin-14 and gastrin-4-like peptides as in the antral G-cells (Rehfeld and Larsson, 1979). The recent production of an antiserum in our laboratory (No. 278) which recognizes gastrin-17 and gastrin-4 with equimolar potency will probably answer the question in a near future.

Although the processing of the gastrin precursor in hypothalamo-hypophyseal neurons resembles that of the main production site (the antral G-cells) there are remarkable exceptions: Pituitary corticotrophs only synthesize non-sulphated gastrins, which apparently are not processed further than gastrin-34 (Rehfeld and Larsson, 1981). Neither do intestinal gastrin cells synthesize gastrin-17 (Larsson and Rehfeld, 1979a), while gastrin in the vagus nerve occurs almost only in heptadecapeptide form (Uvnäs-Wallensten et al., 1977). Finally, gastrin-producing tumours and G-cells in hypergastrinaemic achlorhydria display gross variations in the processing pattern with respect to both proteolysis and sulphation (Andersen and Stadil, 1983). In view of the grave disturbances in sick gastrin cells in the periphery, it would be worthwhile examining the central gastrin processing in diseases in the hypothalamus and pituitary.

Posttranslational Processing of CCK

CCK-containing neurons occur in all regions of the brain but epithalamus and cerebellum (Rehfeld, 1978c; Larsson and

Rehfeld, 1979b). Hypothalamo-hypophyseal neurons display, however a curious species variation. In pig and cat only gastrins are present, while in man, ox and rat the pituitary neurons in addition contain large quantities of CCK. In fact the highest CCK concentrations in the bovine and murine CNS are found in the neurohypophysis (unpublished results). The widespread and abundant occurrence of CCK neurons in mammalian brains makes it almost impossible to disclose whether different CCK processing patterns occur in CNS. However, the possibility definitely exists.

Since CCK-mRNA has not yet been isolated, we only have information about the cerebral CCK biosynthesis from the identification of the various CCK-peptides and from the in vivo chase studies. So far we assume that the CCK peptides originate from a single precursor. The number of different molecular forms identified by chromatography is large (Table I).

TABLE I. Molecular Forms of CCK

Component	K_{av} *(Sephadex G-50)*	*Amino Acids (No.)*	*Brain*	*Gut*
I	0.08	–	+	–
II	0.23	–	+	+
III	0.34	58	+	+
IV	0.51	33/39	+	+
V	0.70	~ 20	–	+
VI	0.92	12	(+)	(+)
VII	1.10	8	+	+
VIII	1.18	~ 6	+	+
IX	1.30	4	+	+

Although sulphated CCK-8 predominates in most CNS regions, careful studies of extracts from large tissue samples with extensive control of possible extraction artefacts and using the different radioimmunoassay principles show that the other forms constitute from 10 to 50% of all the CCKs. The percentage depends on the region examined. Fig. 1 gives an impression of the quantitative relationship between the CCK components in the porcine cerebral cortex as apparent by gel chromatography.

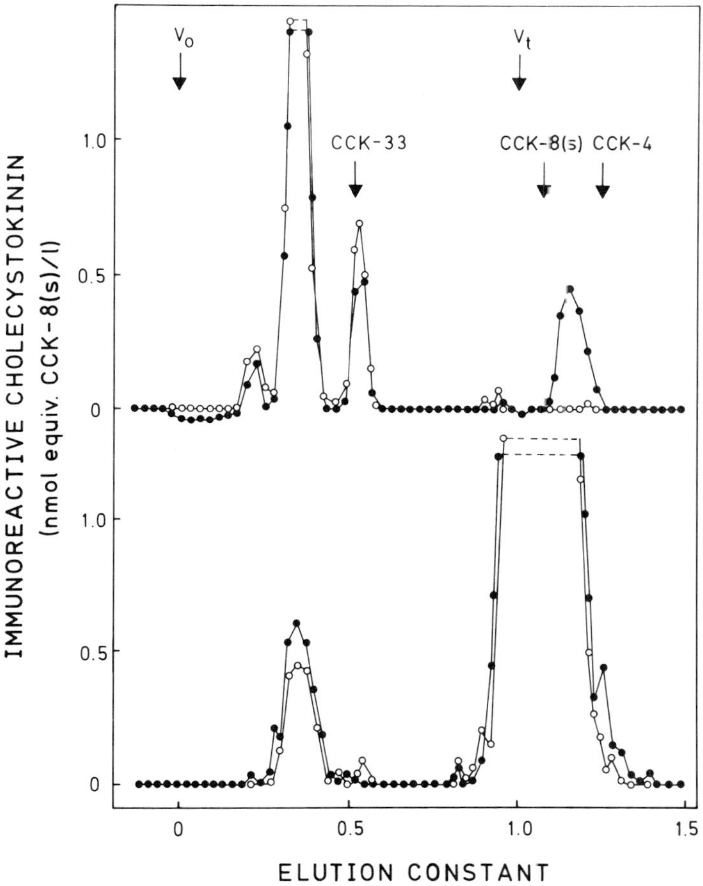

FIGURE 1. Sephadex G-50 gel chromatography of an acid (upper panel) and neutral, boiling water (lower panel) extract of the porcine cerebral cortex. The elutions were monitored by a radioimmunoassay specific for the C-terminal 30-33 sequence (Ab. 2609, ●—●) and a radioimmunoassay specific for sequence 25-30 of CCK-33 (Ab. 4698, O—O).

Obviously, there are substantial amounts of CCK components larger than CCK-8. Notably, extracts of human cerebral cotex display an almost similar CCK pattern by gel chromatography (Rehfeld, 1978c and unpublished results). Fig. 2 further illustrates the heterogeneity with emphasis on the small molecular forms of CCK. When the immunoreactive fractions eluted after the salt peak (V_t) on the Sephadex G-50 columns (Fig.1) are applied to DE-32 cellulose anion exchange chromatography, a considerable microheterogeneity of both the CCK-8- and CCK-4-

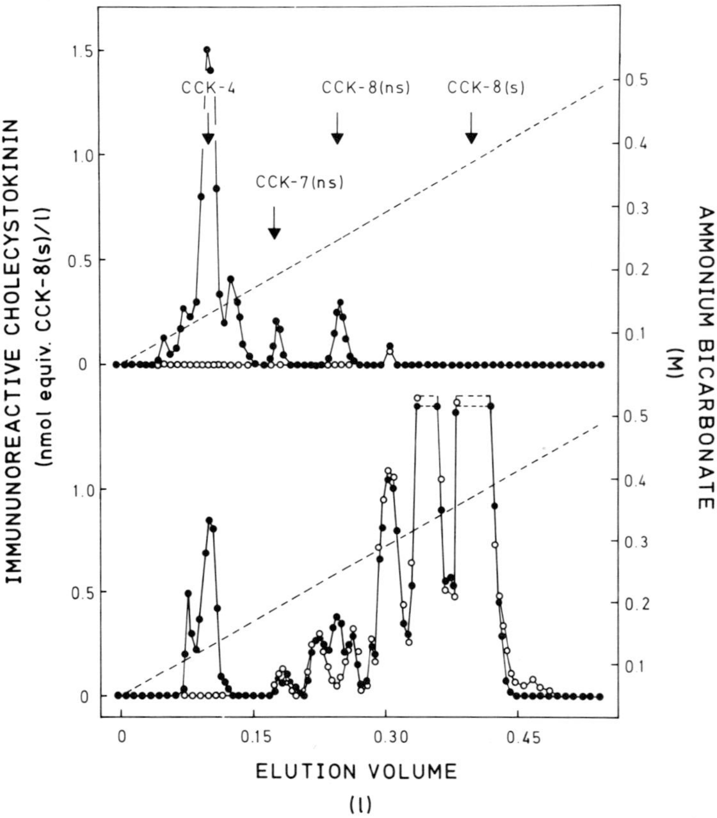

FIGURE 2. DE-32 cellulose gradient elution ion exchange chromatography of the fractions eluted after V_t (Fig. 1) of the acid (upper panel) and neutral (lower panel) cerebrocortical extracts. The elutions were monitored as described in the legend to Fig. 1.

like peptides appears.

The findings justify the conclusion that the posttranslational processing of CCK in the brain involves a multitude of enzymatic modifications. Although a clear-cut CCK precursor-product relationship so far has been shown only for sulphated CCK-8 (Goltermann et al., 1980a), our recent findings (Figs. 1 and 2) corroborate the suggestion that CCK-8 is not the only active end product of the CCK neurons (Rehfeld, 1981).

ACKNOWLEDGMENTS

This review is based on results obtained by an enjoyable collaboration with E. Boel, P.C. Emson, N. Goltermann, C. Grimmelikhuijzen, H.F. Hansen, T. Hökfelt, L.-I. Larsson, K. Marcker, P.D. Marley, J.S. Morley, F. & K. Norris, F. Sundler, K. Uvnäs-Wallensten and J. Vuust. The skillful secretarial assistance of L. Mygil is gratefully acknowledged.

REFERENCES

1. Andersen, B.N., and Stadil, F., *Regul. Peptides*, in press.
2. Boel, E., Vuust, J., Norris, F., Norris, K., Wind, A., Rehfeld, J.F., and Marcker, K.A., *Proc. Natn. Acad. Sci. 80*, in press (1983).
3. Dockray, G.J., *Nature 264*, 568 (1976).
4. Goltermann, N.R., Rehfeld, J.F., and Petersen, H.R., *J. Biol. Chem. 255*, 6181 (1980a).
5. Goltermann, N.R., Rehfeld, J.F., and Petersen, H.R., *J. Neurochem. 35*, 479 (1980b).
6. Grimmelikhuijzen, C., Sundler, F., and Rehfeld, J.F., *Histochemistry 69*, 61 (1980).
7. Larsson, L.-I., and Rehfeld, J.F., *Nature 269*, 335 (1977).
8. Larsson, L.-I., and Rehfeld, J.F., *Nature 277*, 575 (1979a).
9. Larsson, L.-I., and Rehfeld, J.F., *Brain Res. 165*, 201 (1979b).
10. Larsson, L.-I., and Rehfeld, J.F., *Science 213*, 768 (1981).
11. Marley, P.D., and Rehfeld, J.F., *J. Neurochem.*, in press (1983).
12. Noyes, B.E., Mevarech, M., Stein, R., and Agarwal, K., *Proc. Natn. Acad. Sci. USA 76*, 1770 (1979).
13. Rehfeld, J.F., *Acta Toxicol. Pharm. 24*, 44 (1977).
14. Rehfeld, J.F., *Nature 271*, 771 (1978a).
15. Rehfeld, J.F., *J. Biol. Chem. 253*, 4016 (1978b).
16. Rehfeld, J.F., *J. Biol. Chem. 253*, 4022 (1978c).
17. Rehfeld, J.F., *in "Gut Hormones"* (S.R. Bloom and J.M. Polak, eds.), p. 240. Churchill Livingstone, Edinburgh, (1981).
18. Rehfeld, J.F., and Larsson, L.-I., *Acta Physiol. Scand. 105*, 117 (1979).
19. Rehfeld, J.F., and Larsson, L.-I., *J. Biol. Chem. 256*, 10426 (1981).
20. Rehfeld, J.F., and Morley, J.S., *J. Biochem. Biophys. Meth. 7*, 161 (1983).

21. Rehfeld, J.F., Goltermann, N., Larsson, L.-I., Emson, P.C., and Lee, C.M., *Fed. Proc. 38*, 2325 (1979).
22. Uvnäs-Wallensten, K., Rehfeld, J.F., Larsson, L.-I., and Uvnäs, B., *Proc. Natn. Acad. Sci. USA 74*, 5707 (1977).
23. Vanderhaeghen, J.J., Signeau, J.C., and Gepts, W., *Nature 257*, 604 (1975).
24. Vanderhaeghen, J.J., Lotstra, F., Vandesande, F., and Dierickx, K., *Cell Tissue Res. 221*, 227 (1981).

PROCESSING IN NERVES AND ENDOCRINE CELLS OF PEPTIDES RELATED TO FMRFamide AND MET-enkephalin Arg^6Phe^7

G. J. Dockray
A. Giraud

Physiological Laboratory
University of Liverpool
Liverpool, U.K.

J. R. Reeve, Jr.

CURE, VA Hospital Center
Los Angeles
California, U.S.A.

Possible processing pathways of proenkephalin in adrenal chromaffin cells and central and peripheral (colon) neurones have been studied by examining the patterns of molecular forms reacting with specific antisera to Met-enk Arg^6, Met-enk Arg^6 Gly^7 Leu^8, Met-enk Arg^6Phe^7 and its C-terminal tetrapeptide amide FMRFamide. Antibodies to FMRFamide, which is a molluscan neuropeptide, react with material in vertebrate CNS extracts, but the active factors appear to be a distinct group of peptides not related to the opioid series. The distribution of immunoreactive forms of enkephalin hexa- hepta- and octa- peptides indicates that in central and peripheral neurones there is efficient processing to the octa- and heptapeptides, but in adrenal medulla N-terminally extended forms predominate.

BIOCHEMICAL AND CLINICAL ASPECTS OF
NEUROPEPTIDES: SYNTHESIS, PROCESSING,
AND GENE STRUCTURE

ISBN 0-12-417320-9

INTRODUCTION

In recent years considerable interest has developed in the enkephalin-related peptides of the bovine adrenal medulla, largely because these are thought to be possible precursors or intermediates in the biosynthesis of the enkephalin pentapeptides (10,18). It is established that in adrenal chromaffin cells a variety of different opioid peptides may be produced from a common precursor: in addition to Met- and Leu-enkephalin, there are both C-terminally extended variants eg Met-enk Arg^6Phe^7 (17) and Met-enk $Arg^6Gly^7Leu^8$ (9) and N-terminally extended forms (18). However, the extent to which the pattern of post-translational processing of precursors in the adrenal medulla resembles that in neurones remains uncertain. In this context, two possibilities are of particular interest. First, by analogy with the β-endorphin precursor, pro-opiomelanocortin, both opioid and non-opioid biologically active peptides might be produced from a common precursor. Second, different patterns of opioid peptides might be produced in different cells from the same precursor molecule. We have studied aspects of both questions. With regard to the first possibility we embarked, a little over two years ago, on an examination of the hypothesis that peptides containing the sequence Met-enk Arg^6Phe^7 might be processed by cleavage and C-terminal amidation to give the tetrapeptide, FMRFamide:

Met-enk Arg^6Phe^7	Tyr-Gly-Gly-Phe-Met-Arg-Phe
FMRFamide	Phe-Met-Arg-Phe-NH_2

The tetrapeptide amide was isolated from molluscan ganglia by Price and Greenberg (15), and is known to have potent biological actions on many tissues, including mollusc heart and smooth muscle (6). Moreover, it is of interest that FMRFamide has stimulatory effects on neurones of the medullary reticular formation in the rat brain stem while the same cells are depressed by Met-enk Arg^6Phe^7 (5). Initial studies using antibodies specific for FMRFamide were promising in that immunoreactive material was found in a wide variety of vertebrate neurones and endocrine cells (3). Subsequently, however, the idea of a common biosynthetic origin of FMRF-amide and Met-enk Arg^6Phe^7 has become untenable. Thus it is now clear that in many systems C-terminally amidated amino acids are followed in their precursor molecules by glycine, but in the case of proenkephalin, Met-enk Arg^6Phe^7 forms the final C-terminal sequence of the entire precursor and the final Phe residue cannot be therefore followed by the Gly residue needed for C-terminal amidation (2). In addition, direct evidence, some of which is described below, indicates that vertebrate FMRFamide-like immunoreactive material is likely to be due to substances which share only a dipeptide amide with the molluscan tetrapeptide. In contrast, the results of studies on the second possibility mentioned above indicate that the proenkephalin precursor may indeed be processed by different pathways in different cells. To examine this question we have used antisera specific for the C-terminal portions of the hexapeptide, Met-enk Arg^6 (7), the heptapeptide Met-enk Arg^6Phe^7 and octapeptide Met-enkephalin $Arg^6Gly^7Leu^8$ in radioimmunoassays to characterise the amount and type of immunoreactive material in adrenal, brain and colon extracts.

METHODS

Radioimmunoassay

Antisera to FMRFamide, Met-enk Arg^6, Met-enk Arg^6Phe^7 and Met-enk $Arg^6Gly^7Leu^8$, were obtained by coupling these peptides to thyroglobulin with glutaraldehyde (3,4). In each case glutaraldehyde would be expected to conjugate at the N-terminal amino group and so leave the C-terminus free to promote antibody formation. Radio-labelled ligands (Tyr^1 FMRFamide, or the original hapten) were produced by the chloramine T method using $Na^{125}I$, and were purified on CM-Sephadex (1 x 10 cm) using gradient elution with ammonium acetate to separate labelled and unlabelled peptides and so give a mono-iodinated product of high specific activity. Conventional RIA incubation conditions were used and antibody-bound label separated from free label using dextran- and plasma-coated charcoal. The specificity of each assay was established by comparison of the concentration of standard and test peptides needed for 50% inhibition of binding of the label.

Extracts

Studies of FMRFamide-like immunoreactivity (-LI) have been made on chicken brain and rat spinal cord. Met-Arg^6 -,

Met-enk Arg^6Phe^7 - and Met-enk $Arg^6Gly^7Leu^8$-LI were studied in bovine adrenal medulla, striatum and colon muscle. Abattoir material (chicken and bovine) was transported to the laboratory in ice, rapidly dissected and stored at -40°C prior to extraction. Tissues were extracted by slowly adding small pieces of deeply frozen tissue to vigorously boiling water (final concentration $0.1 g.ml^{-1}$). After boiling for at least 5 min, the extracts were cooled, homogenised, centrifuged and the residue re-extracted with acetic acid 0.5M. Both water and acid extracts were concentrated by adsorption to C18 Sep-pak cartridges (Waters Associates, Milford MS, USA) and elution with 80% methanol (4). The products were lyophilized prior to chromatography.

Chromatography

Extracts of chicken brain and rat spinal cord were purified by ion exchange chromatography on CM-Sephadex (1 x 10 cm, gradient elution with ammonium acetate 0.05M to 0.5M). Immunoreactive peaks were concentrated by Sep-pak cartridges as before and fractionated by gel filtration on Sephadex G50 or G25 (1 x 100 cm, 0.5M acetic acid). The products were further purified by reverse phase HPLC using an Altex system and C18 columns (5 x 25 mm Techsil 5μ C-18, Vydac C-18 or Z-module C-18 cartridge from Millipore) or on Spherisorb 5μ phenyl (5 x 25mm). Bovine tissue extracts were applied directly to Sephadex G50 (1 x 100 cm, 0.5M acetic acid). Appropriate dilutions of the column eluates were added to radioimmunoassay tubes: in those cases where more than 20 μl was added to the tube the solutions were first lyophilised to prevent non-specific effects in the assay.

RESULTS

FMRFamide Immunoreactivity

RIA Specificity. Two antibodies have been used in routine assays. Both show high specificity for the C-terminus and vary slightly in the extent to which they tolerate amino acid substitutions in other positions. With antibody L134, avian pancreatic polypeptide (APP) which has the C-terminal tetrapeptide $-Arg-His-Arg-Tyr-NH_2$, shows about 2% immunoreactivity relative to FMRFamide; mammalian pancreatic polypeptide-related molecules (PP, NYP and PYY) have either Gln or Pro at position 3 from the C-terminus, and show only very low immunoreactivity (< 0.001). With L155 neither APP nor the mammalian PP-related peptides have more than 0.3% immunoreactivity relative to FMRFamide. The dipeptide -Arg-Phe-NH_2 occurs in γ_1 MSH, but this shows low activity with both antisera, as does the gastrin tetrapeptide Trp-Met-Asp-Phe-NH_2 (Table 1).

TABLE 1. Specificity of FMRFamide assays using two different antisera (L134 and L155). Immunochemical potency of selected peptides (structures given in the single letter notation) is expressed as the ratio of concentrations of standard and test peptide needed for 50% inhibition of binding of label.

Peptide	*L134*	*L155*
FMRFamide	*1.0*	*1.0*
YGGFMRFamide	*1.29*	*1.83*
YGGFMRF-OH	*< 0.001*	*< 0.001*
APP (-RHRYamide)	*0.017*	*0.001*
BPP6 (-RPRYamide)	*< 0.001*	*0.002*
PYY (-RQRYamide)	*< 0.001*	*0.003*
NPY (-RQRYamide)	*0.001*	*0.003*
γ_1MSH (-WDRFamide)	*0.015*	*< 0.001*
CCK8 (-WMDFamide)	*< 0.001*	*< 0.001*

Chicken FMRFamide-like immunoreactivity. Preliminary studies suggested that the concentrations of FMRF-LI in chicken brain might be sufficient for isolation and chemical characterization. In these extracts, four peaks of immunoreactivity (designated I - IV in order of emergence) were separated on CM-Sephadex, and were then separately run on Sephadex G25 (Fig. 1).

The major peak on CM-Sephadex (II, accounting for about 50% of total) was separated on Sephadex G25 into three peaks (IIa,b,c): peak III emerged as a single component on Sephadex G25. In fraction II the major component, IIb, was separated into two further immunoreactive fractions on reverse phase HPLC on Techsil C18, while in the same system IIc eluted as a single immunoreactive peak. Peak IIc and one of the pair of IIb components (Fig. 2) have been obtained in a pure form suitable for structural analysis.

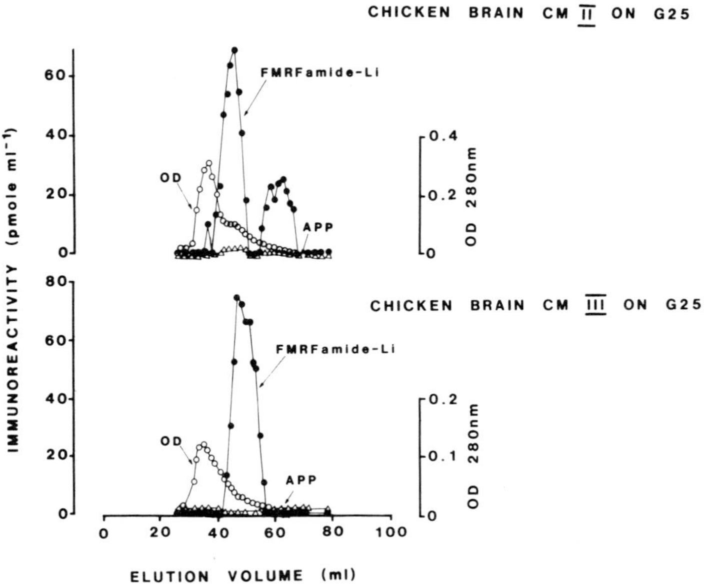

FIGURE 1. Separation on Sephadex G25 of FMRFamide-LI (●——●) corresponding to peaks II and III (the major components) obtained from CM-Sephadex. A_{280} (o——o); avian pancreatic polypeptide-LI (APP, Δ) was undetectable.

Digestion with trypsin of both IIb material and IIc yielded Phe-NH_2 which was identified by A_{212} following isocratic HPLC on Vydac C-18 (3% acetonitrile) where it had a retention time of 14.8 min. Since the penultimate residue must be either Arg or Lys, and since only Arg is tolerated by the present FMRFamide antisera (4), the C-terminal dipeptide must be -Arg-Phe-NH_2. The amino acid composition of peak IIc (in collaboration with Dr. J. Shively) indicated in addition to Arg (1) and Phe (1) the presence of Leu (2) and Pro (1) compatible with a pentapeptide structure. It is of interest that following tryptic digestion there is no increase in immunoreactivity with antiserum specific for C-terminus of Met-enk Arg^6 indicating that FMRFamide-LI in the chicken extracts is not related to the opioid series of peptides.

Preliminary studies suggest that chicken IIc has similar biological actions to FMRFamide in rat. Thus both peptides excite the same medullary reticular formation neurones. In addition, on intravenous or intracisternal administration both peptides increase arterial blood pressure with similar potency and time course of action.

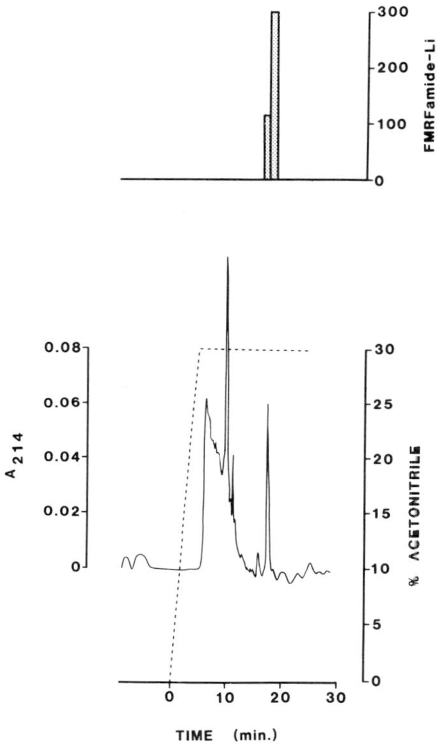

FIGURE 2. Reverse phase HPLC of peak IIb on Vydac C-18 using a gradient of acetonitrile in 0.1% trifluoroacetic acid. The material was previously purified by reverse phase HPLC on Techsil C18 and Spherisorb-phenyl.

Mammalian FMRFamide immunoreactivity. In rat CNS extracts, FMRFamide-LI is found in highest concentrations (10-15 pmol.g^{-1}) in hypothalamus, brain stem and spinal cord (4). Extracts of bovine adrenal medulla, striatum and colon which were rich in enkephalin variants (see below) had little or no FMRFamide-LI. Extracts of rat spinal cord fractionated on CM-Sephadex were resolved into two components with FMRFamide-like activity. The two peaks can also be separated on reverse phase HPLC on Techsil 5μ C-18 when they emerge with retention times greater than that of chicken IIc. As with chicken material, it is noteworthy that digestion with trypsin did not lead to increases in Met-enk Arg^6-LI.

Met-enk Arg^6-LI, Met-enk Arg^6Phe^7 -LI and Met-enk Arg^6Gly^7 Leu^8-LI

Radioimmunoassay Specificity. Antisera specific for the C-terminal regions of Met-enkephalin -Arg^6, -Arg^6 -Phe^7 and -Arg^6 -Gly^7 -Leu^8 each show little or no cross-reactivity with other members of the series. In each case there is a high requirement for a free C-terminus and neither C-terminal extensions, nor C-terminal deletions are tolerated by these antibodies (Table 2). Trypsinisation of synthetic Met-enk Arg^6Phe^7 and Met-enk Arg^6Gly^7 Leu^8 abolishes activity with their respective antibodies and gives 70-100% conversion to immunoreactive Met-enk Arg^6.

Adrenal Medulla. High concentrations of Met-enk Arg^6 Phe^7 - and Met-enk Arg^6 Gly^7 Leu^8-LI were found in extracts of adrenal medulla (1-2 $nmol.g^{-1}$), but in both cases less than 5% of total immunoreactivity eluted in the position of the standard peptide on gel filtration. The majority of the immunoreactivity with both antibodies emerged in the position of higher molecular weight forms (Fig. 3). The heptapeptide immunoreactivity could well correspond to the peptide of 3.6 kDal and the octapeptide-LI with 5.3 and 18.2 kDal peptides previously isolated (8,17,19).

TABLE 2. Relative immunochemical potencies of enkephalin variants with antisera to Met-enk Arg^6 (L167), Met-enk Arg^6Phe^7 (L152) and Met-enk $Arg^6Gly^7Leu^8$ (L189). Potency expressed as the ratio of concentrations of the original peptide and test substances needed for 50% inhibition of binding of label.

Peptide	*L167*	*L152*	*L189*
Met-enk	*<0.001*	*<0.001*	*<0.001*
Met-enk Arg^6	*1.00*	*<0.001*	*<0.001*
Met-enk Arg^6Arg^7	*0.006*	*<0.001*	*<0.001*
Met-enk Arg^6Phe^7	*0.003*	*1.00*	*<0.001*
Met-enk Arg^6Phe^7-NH_2	-	*0.008*	*<0.001*
Met-enk $Arg^6Gly^7Leu^8$	-		*1.00*
Leu-enk Arg^6	*<0.001*	-	*<0.001*

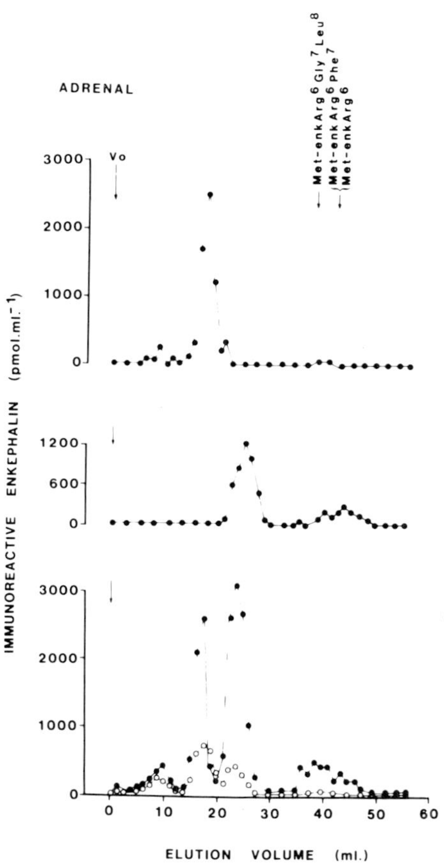

FIGURE 3. Separation on Sephadex G50 of adrenal medulla extracts assayed with L152, L189 and L167 (before (o) and after (●) trypsin).

Concentrations of Met-enk Arg^6-LI were about 10% those of the octa- and heptapeptide variants. But digestion of the column eluates with trypsin produced large increases in Met-enk Arg^6-LI: these were especially striking in the region of the column eluates associated with Met-enk Arg^{6}Phe^7 - and Met-enk Arg^{6}Gly^{7}Leu^8-LI.

Brain and Gut. In sharp contrast to adrenal gland, extracts of striatum and colon contained major peaks of immunoreactive material compatible with Met-enk Arg^{6}Phe^7 and Met-enk Arg^{6}Gly^{7}Leu^8, and only trace amounts of larger forms (Figs. 4 & 5). In both tissues Met-enk Arg^6 was again present only in low concentrations relative to the other forms, but as in the adrenal trypsinization of the column eluates generated Met-enk Arg^6-LI in the position of peaks of Met-enk Arg^{6}Phe^7 and Met-enk Arg^{6}Gly^{7}Leu^8.

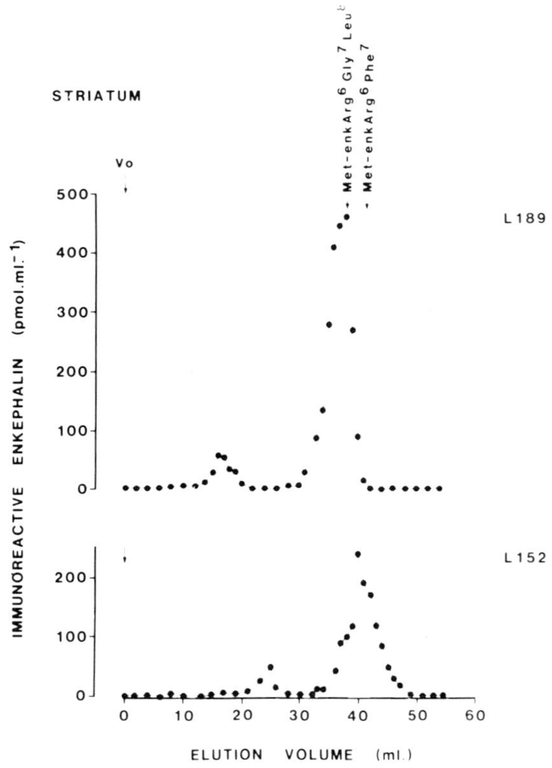

FIGURE 4. Sephadex G50 gel filtration of a bovine striatum extract assayed with antisera L189 and L152.

DISCUSSION

The present results using antisera to FMRFamide point to the conclusion that immunoreactive material in vertebrate CNS extracts is structurally distinct both from the molluscan tetrapeptide, and the opioid heptapeptide Met-enk Arg^{6}Phe^7. The similarities in sequence between FMRFamide and the opioid heptapeptide would therefore seem to be fortuitous, and not the result of a common evolutionary history or biosynthesis. This conclusion is also in agreement with existing data which indicates that the C-terminal Phe residue of Met-enk Arg^{6}Phe^7 (2) is unlikely to be available for amidation by known mechanisms. Although antisera to

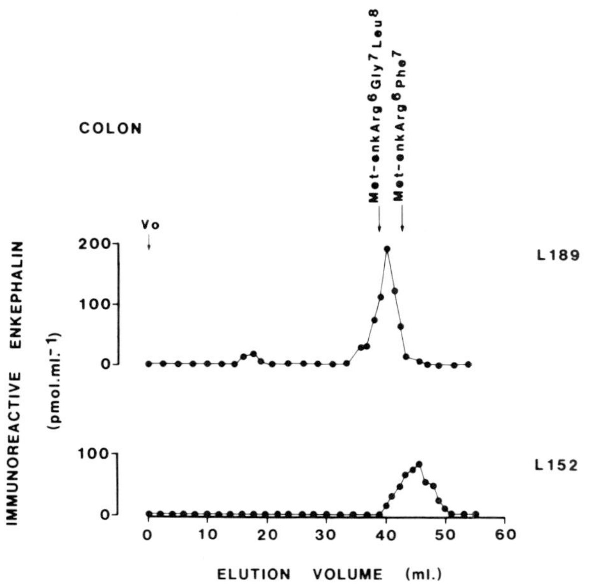

FIGURE 5. Sephadex G50 gel filtration of an extract of bovine colon muscle, assayed with antisera L189 and L152.

FMRFamide react weakly with some known vertebrate neuropeptides and hormones it seems that much of the material revealed by RIA of vertebrate CNS extracts might correspond to a novel series of peptides. The physiological significance of the latter remains to be established.

The different pattern of molecular forms of Met-enk $Arg^6 Phe^7$ and Met-enk $Arg^6Gly^7Leu^8$ in adrenal compared with brain and gut extracts points to differences in the pathways of post-translational processing of proenkephalin in adrenal chromaffin cells on the one hand and in central or peripheral neurones on the other. This conclusion is compatible with observations on the distribution of material reacting with antisera to other opioid peptides first isolated from the adrenal medulla, including synenkephalin and BAM 12 (1,11,13). In neurones it would appear that processing is relatively efficient in producing smaller forms, which have known biological activity (12). In the adrenal, however, processing is less complete and conserves the large forms. It is of interest that the three processing sites involved in the liberation of Met-enk Arg^6Phe^7 and Met-enk $Arg^6Gly^7Leu^8$ all involve cleavage of -Lys-Arg-bonds (14) (Fig. 6).

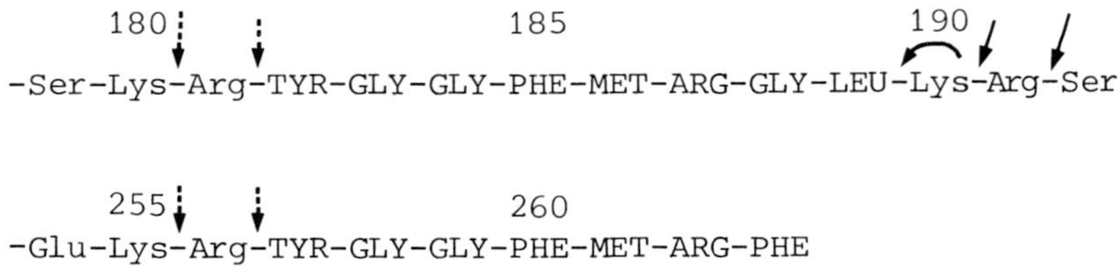

FIGURE 6. Regions of proenkephalin containing the sequences Met-enk Arg⁶Phe⁷, Met-enk Arg⁶Gly⁷Leu⁸. Arrows indicate possible cleavage sites.

Digestion with trypsin of eluates of adrenal columns showed major peaks of Met-enk Arg^6-LI to be generated where Met-enk Arg^6Phe^7 or Met-enk $Arg^6Gly^7Leu^8$-LI occurred. It would seem that processing is relatively efficient at the C-terminus of Met-enk $Arg^6Gly^7Leu^8$, since C-terminally extended forms of this peptide would also have been found after trypsinization. Evidently, in the adrenal medulla there is a mechanism that favours cleavage at the C-terminal side of the octapeptide but not at the immediate N-terminal side of either the octa- or heptapeptides.

A free N-terminal Tyr residue is known to be needed for opioid activity, so that the sequences of Met-enk Arg^6Phe^7 and Met-enk $Arg^6Gly^7Leu^8$ in adrenal medulla are biologically masked. It is possible, of course, that the larger forms are biologically active but act at non-opioid receptors. Alternatively it is possible that conversion of the larger forms to active molecular species occurs after their release by mechanisms of post-secretory processing.

REFERENCES

1. Baird, A., Ling, N., Bohlen, P., Benoit, R., Klepper, R. and Guillemin, R. *Proc. Natl. Acad. Sci. USA. 79,* 2023 (1982).
2. Bradbury, A.F., Finnie, M.D.A. and Smyth, D.G. *Nature, 298,* 686 (1982).
3. Dockray, G.J., Vaillant, C., Williams, R.G. *Nature, 293,*

565 (1981).

4. Dockray, G.J. and Williams, R.G. *Brain Res. 266,* 295 (1983).
5. Gayton, R.J. *Nature, 298,* 275 (1982).
6. Greenberg, M.J. and Price, D.A. *in* "Peptides: integrators and cell tissue function" (F. Bloom, ed), p. 107, Raven Press, New York (1980).
7. Huang, W.-Y., Chang, R.C.C., Kastin, A.J., Coy, D.H. and Schally, A.V. *Proc. Natl. Acad. Sci. USA 76,* 6177 (1979).
8. Jones, B., Shively, J.F., Kilpatrick, D.L., Kojima, K. and Udenfriend, S. *Proc. Natl. Acad. Sci. USA 79,* 1313 (1982).
9. Kilpatrick, D.L., Jones, B.N., Kojima, K. and Udenfriend, S. *Biochem. Biophys. Res. Comm. 103,* 698 (1981).
10. Kojima, K., Kilpatrick, D.L., Stern, A.S., Jones, B.N., Udenfriend, S. *Archi. Biochem. Biophys. 215,* 638 (1982).
11. Liston, D.R., Vanderhaeghen, J.-J. and Rossier, J. *Nature, 302,* 62 (1983).
12. Magnan, J., Paterson, S.J. and Kosterlitz, H.W. *Life Sci. 31,* 1359 (1982).
13. Mizuno, K., Minamino, N., Kangawa, K. and Matsuo, H. *Biochem. Biophys. Res. Comm. 97,* 1283 (1980).
14. Noda, M., Furutanyi, Y., Takahashi, H., Toyosato, M., Hirose, T., Inayama, S., Nakanishi, S. and Numa, S. *Nature, 295,* 202 (1982).
15. Price, D.A. and Greenberg, M.J. *Science, 197,* 670 (1977).
16. Stern, A.S., Lewis, R.V., Kimura, S., Rossier, J., Gerber, L.D., Brink, L., Stein, S. and Udenfriend, S. *Proc. Natl. Acad. Sci. USA 76,* 6680 (1979).
17. Stern, A.S., Lewis, R.V., Kimura, S., Rossier, J., Gerber, L.D., Brink, L., Stein, S. and Udenfriend, S. *Proc. Natl. Acad. Sci. USA, 76,* 6680 (1979).
18. Udenfriend, S. and Kilpatrick, D.L. *Arch. Biochemi. Biophys. 221,* 309 (1983).

BIOSYNTHESIS OF PEPTIDES IN AMPHIBIAN SKIN

Werner Hoffmann
Klaus Richter
Anton Hutticher
Günther Kreil

Institute for Molecular Biology
Austrian Academy of Sciences
Salzburg, Austria

I. INTRODUCTION

Amphibian skin produces a variety of peptides similar or identical to mammalian hormones and neuropeptides (1-3). Upon suitable stimulation, many of these peptides are released as part of a foamy secretion, but their role in this or other processes is still a matter of speculation. It is noteworthy, however, that the quantities of such peptides present in amphibian skin are often far greater than found in any mammalian tissue. For example, the amount of thyrotropin releasing hormone, originally isolated from several hundred thousand ovine or porcine hypothalami could, ironically, have been obtained from the skin of a few dozen frogs like Xenopus laevis (4-6). Large species differences do exist, however, so that a peptide found in abundance in the skin secretion of a particular frog may be virtually absent in a close relative. A partial list of peptides from amphibian skin and their mammalian counterparts is given in Table 1. In several instances it has also been demonstrated that homologous peptides exist in invertebrates.

ISBN 0-12-417320-9

TABLE I. Homologous peptides from amphibian skin and from the gastrointestinal tract and the nervous system of mammals

Amphibian skin	Mammalian tissues
Caerulein Phyllocaerulein	Gastrin Cholecystokinin
Thyrotropin R.H.	Thyrotropin R.H.
Xenopsin	Neurotensin
Bombesin Alytesin Ranatensin etc.	Bombesin
Physalaemin Phyllomedusin Angiotensin	Substance P
Sauvagine	Corticotropin R.H.
Dermorphin	Enkephalin Casomorphin

In the past few years we have been studying the biosynthesis of these frog skin peptides for several reasons. The structure of the precursors of these peptides may yield information about the evolution of mammalian hormones and also lead to new concepts about their stepwise processing to the final product. Moreover, the amphibian skin may represent a suitable model system from which analogies can be drawn for studies on the formation of homologous peptides present in minute quantities in different regions of the brain and other parts of the nervous system (7,8).

II. BIOSYNTHESIS OF CAERULEIN

A decapeptide was isolated from the skin of the Australian frog Hyla caerulea which has the same carboxy-terminal sequence as cholecystokinin and gastrin (9). This peptide, termed caerulein, was subsequently detected in several other species including *Xenopus laevis* (10). In its physiological properties, caerulein resembles cholecystokinin and gastrin and it has indeed been suggested that this amphibian peptide may represent a common ancestor of these hormones (11). Besides their role in the gastrointestinal tract, fragments of cholecystokinin and gastrin have been detected in various parts of the nervous system of mammals. The structure of the precursor of porcine gastrin has been elucidated by cloning the corresponding cDNA (12) but nothing is presently known about the formation of cholecystokinin. We have decided to investigate the biosynthesis of caerulein by a similar approach as used for gastrin.

Starting from the Trp-Met sequence of caerulein, an octanucleotide can be predicted for the mRNA and the corresponding cDNA and the latter can be extended if one allows for dG-U pairs in DNA-RNA-hybrids (see Fig.1). As described in a recent publication (13), the decamer d(AGTCCATCCA) has been synthesized and used as primer in the reverse transcription of poly(A)-containing RNA prepared from the skin of *X. laevis.* The mixture of radioactive cDNA thus obtained was used for colony hybridization.

Using oligo(dT) as primer, poly(A) containing RNA was transcribed into cDNA. Double stranded cDNA was inserted into the unique PstI site of the plasmid pUC8 (14) via poly(dG)-poly(dC) extensions. From the ampicillin-resistent transformants, 300 were selected at random and numbered pUF1-300 (13). In a second experiment, cDNA-RNA hybrids (15) were cloned into the same plasmid and 1400 clones were selected. In both cDNA libraries, close to 10% of the clones hybridized with the radioactive cDNA and many of these clones do indeed contain inserts derived from the mRNA of the caerulein precursor.

```
                 SO3H
                  |
a)   pGlu-Gln-Asp-Tyr-Thr-Gly-Trp-Met-Asp-Phe-NH2
                          ______________________
b)       CA(A/G) GA(T/C) UA(U/C) ACX GGX UGG AUG GA(U/C) UU(U/C) ...3'
c)       cDNA synthesis       ACC TAC CTG A......5'
         <--------------------------------------
```

FIGURE 1. Structure of caerulein and its mRNA.
a) Amino acid sequence of caerulein. The region common to mammalian gastrin and cholecystokinin is underlined
b) Partial sequence of the corresponding mRNA
c) Predicted decadeoxynucleotide used as primer for the synthesis of radioactive cDNA

Originally, the three clones pUF37, pUF48 and pUF262 with a common HindIII site in the insert were sequenced (13). Each of these very similar clones contains a poly(A) tail. The nucleotide sequence of pUF37 with the longest insert is shown in Fig. 2

```
  ACCTCAACAACGAGAAGCCAATGACGAACGTCGCTTTGCTGATGGA
...ProGlnGlnArgGluAlaAsnAspGluArgArgPheAlaAspGly

CAACAAGACTACACAGGTTGGATGGATTTTGGCCGCCGTGATGATGAA
GlnGlnAspTyrThrGlyTrpMetAspPheGlyArgArgAspAspGlu

GATGATGTAAACGAACGAGATGTCCGAGGATTTGGCTCTTTCCTAGGT
AspAspValAsnGluArgAspValArgGlyPheGlySerPheLeuGly

AAAGCTTTAAAGGCTGCTTTAAAAATTGGTGCAAATGCGCTGGGAGGA
LysAlaLeuLysAlaAlaLeuLysIleGlyAlaAsnAlaLeuGlyGly

TCACCTCAACAACGAGAAGCCAATGACGAACGTCGCTTTGCTGATGGA
SerProGlnGlnArgGluAlaAsnAspGluArgArgPheAlaAspGly

CAACAAGACTACACAGGTTGGATGGATTTTGGCCGCCGCAATGGTGAA
GlnGlnAspTyrThrGlyTrpMetAspPheGlyArgArgAsnGlyGlu

GATGATTAATATTCTTCTTGAAAAC(6o bases)poly(A)
AspAsp///
```

FIGURE 2. Nucleotide sequence of the insert of clone pUF37 and amino acid sequence of the open reading frame. Clone pUF48 starts at the A marked (●) and differs by only two point mutations near the 3'-end. Caerulein sequences are underlined, pairs of arginine residues are emphasized. The direct repeat TGATG is marked above the sequence.

The insert of clone pUF37 codes for a polypeptide of 97 amino acids where the decapeptide caerulein occurs twice. This represents the carboxy-terminal part of a larger precursor for this frog skin peptide. In this sequence, the first and the last 33 amino acids are almost identical. More recent results on the sequence of different clones coding for the caerulein precursor suggest the following overall structure: an amino-terminal region of about 30 residues which includes a signal peptide; a stretch of 64 amino acids, separated by the dipeptide Val-Asn from the very similar segment which represents the last 64 amino acids of clone pUF37. The cDNA sequences obtained for the caerulein precursor thus demonstrate that the corresponding gene must have been formed through duplication of a DNS segment. The close similarity indicates that this duplication must have occurred only recently on an evolutionary time scale or that through gene conversion the high degree of homology has been maintained.

In another clone, pUF262, a variant of the caerulein precursor has been detected. This clone contains an insert with a deletion of 45 bp which codes for one of the caerulein copies (13). A second clone with exactly the same delection, pUF11/54 has recently been sequenced so that the possibility of a cloning artefact becomes extremely remote. Interestingly, we have also found two clones in our cDNA libraries, numbered pUF11/9 and pUF11/61, which contain, as compared to pUF37, an insertion that exactly matches the deletion in pUF262 and pUF11/54. We consider these variant caerulein precursors to be the result of an unequal crossing over at the direct repeat TGATG which flanks the segment deleted or inserted in the sequence of the different cDNA clones (see Fig. 3). Besides the mRNA represented in clone pUF37, which codes for a caerulein precursor containing two copies of this peptide, others must be present in skin of Xenopus l. which code for precursors where caerulein occurs one or three times. It could be speculated that we have detected different stages in the evolution of a polyprotein precursor in one tissue. Starting from a hypothetical form with only one copy of the DNA segment coding for the fragment of 64 amino acids mentioned above, through its duplication the

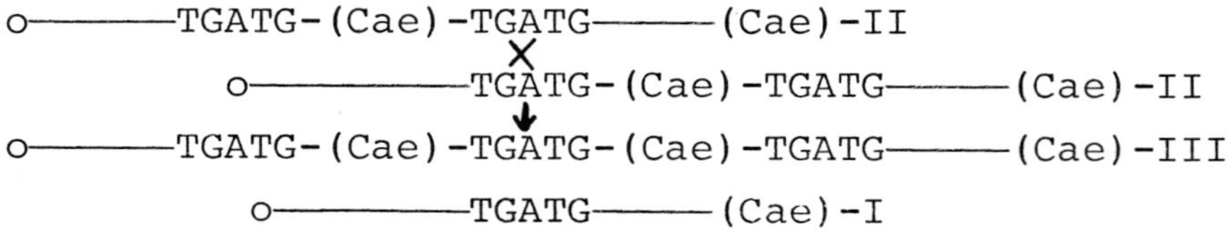

FIGURE 3. Postulated crossing over to create genes for preprocaeruleins I and III. (Cae) = segment coding for caerulein. The repeat TGATG is underlined in Fig. 2.

gene represented in clone pUF37 was generated. Through unequal crossing over the two other forms were created, of which one already contains three copies of the caerulein sequence. We have termed the different variants preprocaerulein I, II and III to indicate the number of caerulein copies present in the precursors. Additional clones with still larger inserts, which may code for other variants of this precursor are under investigation. Beyond the common peptide Gly-Trp-Met-Asp-Phe-Gly-Arg-Arg there are no similarities between the precursor for porcine gastrin (12) and the caerulein precursors.

From the amino acid sequence of the different variants of the caerulein precursors it is evident that the liberation of caerulein must involve several processing steps. The typical prohormone-hormone conversion mechanism proceeds via cleavage at pairs of lysine/arginine residues and their subsequent excision by a carboxypeptidase B-like activity (16). In case of the caerulein precursor this would yield an intermediate form of the final product with an extension at the amino end and an additional glycine residue at the COOH-terminus. The latter represents the recognition site for the formation of a terminal amide, as suggested for the melittin precursor (17) and since found in several other precursors of amidated peptide hormones (18-23). The further maturation of the resulting amidated peptide with the NH_2-terminal sequence Phe-Ala-Asp-Gly or, in one copy of preprocaerulein III, Asp-Gly, apparently occurs by a mechanism first encountered in studies on the biosynthesis of melittin. The activation of promelittin proceeds by stepwise cleavage of dipeptides by a dipeptidylpeptidase which primarily hydrolyzes after proline and alanine residues (24). Sequential removal of dipeptides terminating with alanine also represents the last step in the liberation of yeast alpha-mating factor from its precursor (25,26). Cleavage of the amino-terminal extensions from these postulated intermediates in the maturation of caerulein could also proceed in a stepwise fashion. In skin secretion of Xenopus laevis, a dipeptidylaminopeptidase activity has been detected, which readily cleaves synthetic substrates like Ala-Pro-pNA and Ala-Ala--pNA (pNA = paranitroanilide). We have synthesized the three tetrapeptides Ala-Pro-Ala-Ala, Ala-Ala-Ala-Ala, and Ala-Gly-Ala-Ala, in which the COOH-terminal alanine is radioactive. The relative rate of hydrolysis of these tetrapeptides by partially purified dipeptidylpeptidase from skin secretion was 100:23:15. It is thus quite likely that stepwise cleavage of dipeptides also plays a role in the liberation of caerulein from these precursors.

III. PRECURSORS FOR OTHER FROG SKIN PEPTIDES

During the search for caerulein clones in our cDNA libraries, we have several times encountered clones with inserts coding for an unknown polypeptide. The largest insert sequenced so far could code for 66 amino acids (27) starting precisely with a codon for methionine. The first 20 amino acids which could represent the signal peptide show homology with the signal peptide of preprocaerulein. If the insert in this clone pUF38 really codes for a complete primary translation product, this would be the smallest prepro-peptide yet encountered. In this predicted polypeptide, the sequence ..Lys-Arg-Tyr-Val-Arg-Met-Ala-Ser-Lys-Ala-Gly-Ala-Ile-Ala-Gly-Lys-Ile-Ala-Lys-Val-Ala-Leu-Lys-Ala-Leu-Gly-Arg-Arg..is present. Processing at pairs of basic residues and amidation with concomitant loss of the terminal glycine residue would then yield a peptide of 24 amino acids with an unusual amino acid composition. When arranged into an alpha helix, this hypothetical peptide can form a perfect amphipathic structure. Several peptides with a high affinity for cell membranes have been shown to form such amphipathic helices, such as delta hemolysin (29), cecropin (30), and melittin and similar synthetic peptides (28). We are currently trying to isolate this peptide from frog skin secretions.

Some of the clones which have been sequenced contain inserts which could code for cytoskeletal proteins. Using similar approaches, we are also screening our cDNA libraries for the precursors of other skin peptides like thyrotropin releasing hormone and peptides related to substance P.

ACKNOWLEDGEMENTS

This work was supported by a grant from the Politzer Stiftung of the Austrian Academy of Sciences.

REFERENCES

1. Erspamer, V., Ann. Rev. Pharmacol. 11, 327-350 (1971)
2. Bertaccini, G., in Glass, G.B.J. (ed.) Gastrointestinal Hormones: Peptides of the Amphibian Skin Active on the Gut (Raven Press, N.Y.) pp. 315-341 (1980).
3. Erspamer, V., and Melchiorri, P., Trends in Pharmacol. Sci. 1980, p. 391-395.
4. Guillemin, R., Science 202, 390-395 (1978)
5. Jackson, I.M.D., and Reichlin, S., Science 198, 414-415 (1977).
6. Bennet, G.W., Balls, M., Clothier, R.H., Marsden, M.A., Robinson, G., and Wemss-Holden, G.D., Cell Biol. Int. Rep. 5, 151-158.
7. Snyder, S.H., Science 209, 976-983 (1980).
8. Hökfelt, T., Hohansson, O., Ljungdahl, A., Lundberg, J.M., and Schultzberg, M., Nature 284, 515-521 (1980).
9. Anastasi,A., Erspamer, V., and Endean, R., Arch. Biochem. Biophys. 125, 57-68 (1968).
10. Anastasi, A., Bertaccini, G., Cei, J.M., DeCaro, G., Erspamer, V., Impicciatore, M., and Roseghini, M., Brit. J. Pharmacol. 38, 221-228 (1970)
11. Rehfeld, J.F., Americ.J.Physiol. 240, G255-G266 (1981).
12. Yoo, O.J., Powell, C.T., and Agarwal, K.L., Proc. Natl. Acad. Sci. USA 79, 1049-1053 (1982)
13. Hoffmann, W., Bach, T.C., Seliger, H. and Kreil, G., EMBO J. 2, 111-114 (1983).

14. Vieira, J., and Messing, J., Gene 19, 259-268 (1982).
15. Roninson, I.B., and Ingram, V.M., Proc. Natl. Acad. Sci. USA 78, 4782-4785 (1981).
16. Steiner, D.F., Kemmler, W., Tager, H.S., and Peterson, J.D., Fed. Proc. 33, 2105-2115 (1974).
17. Suchanek, G., and Kreil, G., Proc. Natl. Acad. Sci. USA 74, 975-978 (1977).
18. Amara, S.G., David, D.N., Rosenfeld, M.G., Ross, B.A., and Evans, R.M., Proc. Natl. Acad. Sci. USA 77, 4444-4448 (1980).
19. Land, H., Schütz, G., Schmale, H., and Richter, D., Nature 295, 299-303 (1982).
20. Land, H., Grez, M., Ruppert, S., Schmale, H., Rehbein, M., Richter, D., and Schütz, G., Nature 302, 342-344 (1983).
21. Furutani, Y., Morimoto, Y., Shibahara, S., Noda, M., Takahashi, T., Hirose, T., Asai, M., Inayama, S., Hayashida, H., Miyata, T., and Numa, S., Nature 301, 537-540 (1983).
22. Scheller, N.G., Jackson, J.F., McAllister, L.B., Rothman, B.S., Mayeri, E., and Axel, R., Cell 32, 7-22 (1983).
23. Seidah, N.G., Rochemont, J., Hamelin, J., Benjannet, S., and Chrétien, M., Biochem. Biophys. Res. Commun. 102, 423-427 (1979).
24. Kreil, G., Haiml, L., and Suchanek, G., Eur. J. Biochem. 111, 49-58 (1980).
25. Kurjan, J., and Herskowitz, I., Cell 30, 933-943 (1982).
26. Julius, D., Blair, L., Brake, A., Sprague, G., and Thorner, J., Cell 32, 839-852 (1983).
27. Hoffmann, W., Richter, K., and Kreil, G., EMBO J. 2, 711-714 (1983).
28. DeGrado, W.F., Kézdy, F.J., and Kaiser, E.T. J.Am.Chem.Soc. 103, 679-681 (1981).
29. Freer, J.H., and Birkbeck, T.H., J. Theor.Biol. 94, 535-540 (1982).
30. Merrifield, R.B., Vizioli, L.D., and Boman, H. G., Biochemistry 21, 5020-5031 (1982).

PROCESSING SIGNALS OF THE VASOPRESSIN AND OXYTOCIN PRECURSORS

Hartwig Schmale
Dietmar Richter

Institut für Physiologische Chemie
Abteilung Zellbiochemie
Universität Hamburg
Hamburg, FRG

The peptide hormones vasopressin and oxytocin are expressed as larger precursors - polyproteins - which are processed into the functional peptides by converting enzymes.

INTRODUCTION

Many neuropeptides are synthesized as much larger precursors called polyproteins (Richter, 1983). Like other secretory proteins their synthesis occurs on the rough endoplasmic reticulum and their packaging, processing and transport in secretory vesicles (Steiner et al., 1983).

The first hormones proposed to be synthesized as precursors were vasopressin and oxytocin. In 1964 Sachs and co-workers suggested that vasopressin was initially synthesized as inactive peptide which was converted into the active hormone (Sachs and Takabatake, 1964). Solid information on hormone precursors, however, came only years later from the studies of pro-insulin (Steiner et al., 1983). Since then many other peptide precursors have been identified and sequenced.

BIOCHEMICAL AND CLINICAL ASPECTS OF
NEUROPEPTIDES: SYNTHESIS, PROCESSING,
AND GENE STRUCTURE

ISBN 0-12-417320-9

With the discovery of the ACTH precursor (reviewed in Herbert, 1982) it became clear that some of these precursors were rather complex in their structural organization containing more than one functional peptide. Figure 1 shows examples of these precursors. In analogy to composite viral precursors obtained after virus infection and made up of several distinct viral peptides these neuropeptide precursors have been called cellular polyproteins (reviewed in Richter, 1983).

As indicated in Figure 1 the biological entities within the hormone precursors are bounded by basic amino acids generally assumed to serve as processing signals directing the conversion of the biologically inactive polyproteins into the active peptides. In the case of the vasopressin and oxytocin precursors this process may occur in the Golgi apparatus and in the neurosecretory vesicles during their transport from the hypothalamus to the posterior pituitary (Brownstein <u>et al.</u>, 1980).

In the following article the processing signals involved in the maturation process of vasopressin and oxytocin will be discussed.

PROCESSING AND MODIFICATION SIGNALS

Inspection of the vasopressin precursor sequences from rat and calf indicates that the processing signals have been well conserved. As with many other hormone precursors signal peptide and prohormone are separated by a small neutral amino acid, in this case an alanine residue:

AVP RAT	THR SER ALA CYS	ACC UCU GCC UGC
AVP CALF	————————	—————U——
OT RAT	————————	———C————
OT CALF	————————	———C————

The corresponding nucleotide sequence of the four precursors shows only one change in the third base also reflecting the high degree of conservation.

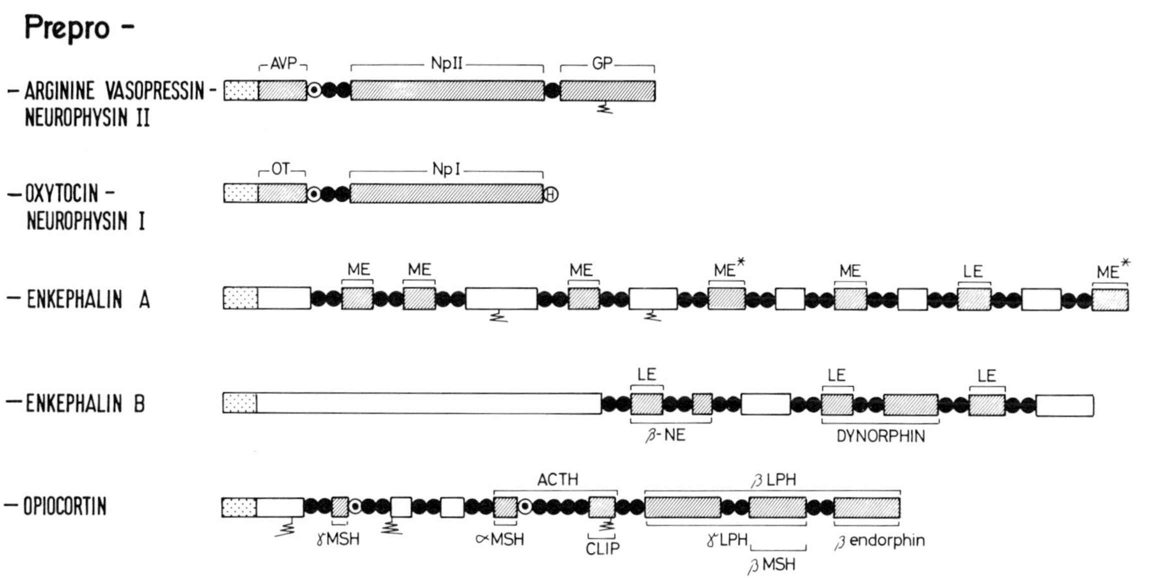

Figure 1. Structures of precursors of some neuropeptide hormones. Strippled bars, pre-sequence; hatched bars, in vivo isolated neuropeptides; open bars, "spacer" sequences (the bars are not drawn to scale); filled circles, basic amino acids; open circles with dots, glycine residues; zig-zag lines, glycosylation site; H, histidine residue. AVP, arginine vasopressin; NpII, bovine neurophysin-II; OT, oxytocin; NpI, bovine neurophysin-I;ME, methionine enkephalin; ME, C-terminally extended methionine enkephalin; LE, leucine enkephalin; ß-NE, ß-neo-endorphin; ß- or γ-LPH, ß- or γ-lipotropin; MSH, melanotropin-stimulating hormone; CLIP, corticoptropin-like intermediate-lobe peptide; ACTH, adrenocorticotropin. Taken from Richter, 1983.*

Other processing signals present in the two precursors are the triplet Gly-Lys-Arg separating the hormones from their carrier proteins. In all cases where information is available on precursor sequences to amidated peptide hormones a glycine residue immediately follows the functional peptide sequence (Table 1, reviewed in Richter, 1983). Except for melittin the glycine is followed by a typical prohormone processing signal, a pair of basic amino acids, or less frequently by a single basic amino acid residue as in the case of prepro-GRF (Gubler et al., 1983).

TABLE I. Partial Amino Acid Sequence of Precursors to Oligo-Peptides Amidated at the Carboxy Terminus

MELITTIN	-ARG-LYS-ARG-GLN-GLN-GLY-COOH
α-MSH	-TRP-GLY-LYS-PRO-VAL-GLY-LYS-LYS-
γ-MSH	-ARG-TRP-ASP-ARG-PHE-GLY-ARG-ARG-
VASOPRESSIN	-ASN-CYS-PRO-ARG-GLY-GLY-LYS-ARG-
OXYTOCIN	-ASN-CYS-PRO-LEU-GLY-GLY-LYS ARG-
CAERULEIN	-GLY-TRP-MET-ASP-PHE-GLY-ARG-ARG-
GASTRIN	-GLY-TRP-MET-ASP-PHE-GLY-ARG-ARG-
CALCITONIN	-GLY-VAL-GLY-ALA-PRO-GLY-LYS-LYS-
HUMAN JOINING PEPTIDE	-PRO-GLY-PRO-ARG-GLU-GLY-LYS-ARG-
EGGLAYING HORMONE	-ARG-LEU-LEU-GLU-LYS-GLY-LYS-ARG-
CORTICOTROPIN RELEASING FACTOR	-LEU-LEU-ASP-ILE-ALA-GLY-LYS-COOH
GROWTH HORMONE RELEASING FACTOR	-ALA-ARG-ALA-ARG-LEU-GLY-ARG-GLN-

Recently a pituitary enzyme has been characterized involved in the amidation reaction which catalyses the oxidative transamidation of a model tripeptide only when a glycine is present with a free carboxy group (Bradbury et al., 1982). From these data it appears that the other processing signals, e.g. Lys-Arg etc. have to be cleaved off before amidation can take place.

The single arginine residue adjacent to the

glycine in the GRF precursor is probably the exception rather than the rule as a processing signal. Other examples where precursors contain single basic residue processing sites are preprosomatostatin, preprovasopressin or CCK.

Preprooxytocin lacks the glycoprotein moiety; instead its neurophysin is extended by an extra arginine (rat preprooxytocin) (Ivell and Richter, 1983). The corresponding bovine precursor carries in this position a histidine (Land et al., 1983) with a single base change in the second position of the codon:

	GLU	ARG
OT RAT	GAC	CGC
OT CALF	C——	—A–
	GLN	HIS

Similar modified processing signals have been reported for pro-insulin (Robbins et al., 1981) and pro-albumin (Abdo et al., 1981; Brennan and Carrell, 1978) where the basic amino acid arginine is mutated to a glutamine (Christchurch pro-albumin) or to a histidine (pro-albumin Lille). While in the cases of pro-albumin precursors processing is greatly impaired, this is not the case for the prooxytocin.

The single arginine residue at the C terminus of the rat oxytocin precursor may represent the rudiment of a once present peptide extending downstream of the basic amino acid. This peptide could be of the same ancestral origin as the glycoprotein moiety of the vasopressin precursor. In the latter case, the glycoprotein had been preserved suggesting it to be of functional significance. This can also be inferred from the well conserved sequence of the glycoprotein from various species. Although the glycoprotein has been detected in hypothalamic and posterior pituitary tissues its function remains to be established.

A posssible modification site can be postulated within the neurophysin sequence at the tyrosine residue (no 61) which could be a potential site for sulfation. This speculation is based on sequence comparison with sulfated peptides such as gastrin where typically acidic amino acids are clustered upstream of the tyrosine:

Gastrin	-Glu-Glu-Glu-Ala-Tyr-
AVP rat	-Gln-Glu-Glu-Asn-Tyr-
AVP calf	————————————
OT rat	————————————
OT calf	————————————

So far neither sulfated preprovasopressin or prepro-oxytocin nor sulfated neurophysins have been reported. However isolation conditions may have hindered the detection of the sulfated products (Huttner, 1982). Although there is no evidence yet for a sulfated neurophysin the position of the tyrosine as a possible sulfation target would have implications for the conformation of the precursors. As reported, sulfation of proteins may be a widespread phenomenon although like phosphorylation its role in the cell is still unknown (Huttner, 1982).

SEQUENCE OF PROCESSING EVENTS

A still unresolved problem is the order in which the peptides are released from the precursors. The processing model proposed in Figure 2 is based on studies of Xenopus laevis oocytes programmed with hypothalamic mRNA (Schmale and Richter, 1980). There, the first cleavage of the vasopressin prohormone occurs between neurophysin and the glycoprotein which are separated by a single arginine residue. The second cleavage step between hormone and carrier protein has not been observed. This would suggest that the converting enzymes of the oocyte recognize a single arginine, but not, however, a pair of basic amino acids. It has been similarly shown that oocytes are unable to cleave the paired basic residues which are the cleavage signals for converting proinsulin (Rapoport et al., 1978).

In the hypothalamic system, alternatively the first cleavage step could occur between hormone and neurophysin. The experiments with the oocytes suggest that there are sets of converting enzymes with different specificities. Whether unmasking of the respective cleavage sites is also important for the order of sequence events cannot be entirely excluded.

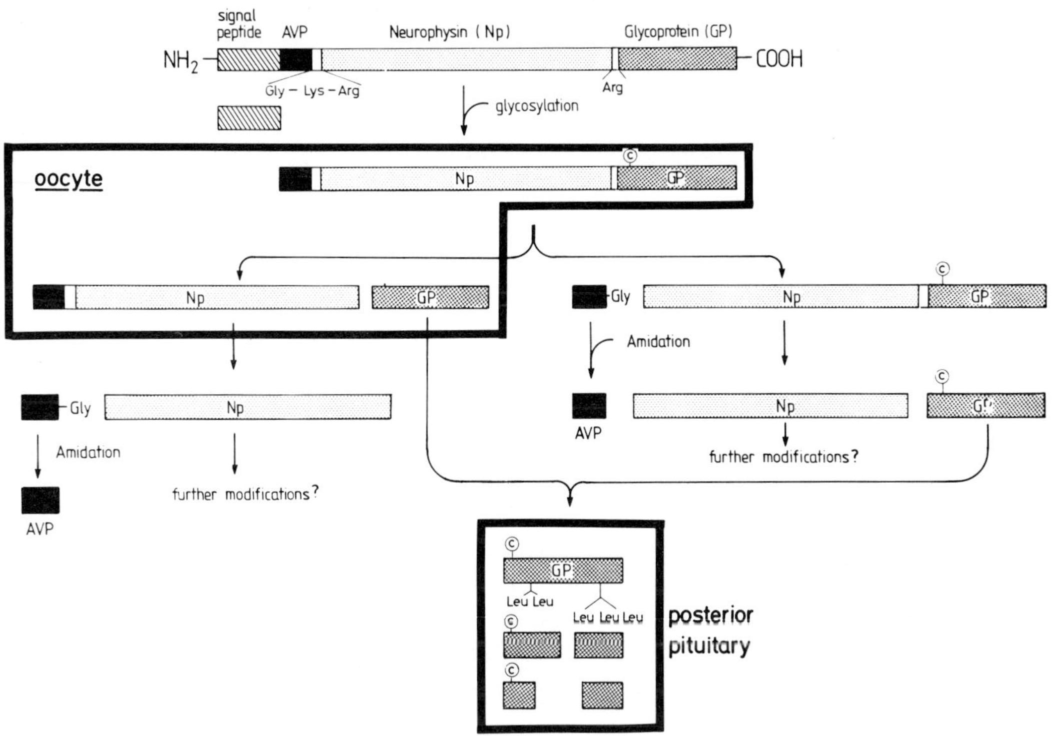

Figure 2. Possible pathways of the vasopressin precursor. The left part (framed) shows processing steps observed in oocytes programmed with hypothalamic mRNA. Alternative processing steps are indicated in the right panel. Processed fragments of the glycoprotein has been found in the posterior pituitary.

REFERENCES

Abdo, Y., Rousseaux, J., and Dautrevaux, M., *FEBS Lett. 131,* 286 (1981).

Bradbury, A.F., Finnie, M.D.A. and Smyth, D.G., *Nature 298,* 686 (1982).

Brennan, S.O. and Carrell, R.W., *Nature 274,* 908 (1978).

Brownstein, M.J., Russell, J.T. and Gainer, H., *Science 207,* 373 (1980).

Gubler, U., Monahan, J.J., Lomedico, P.T., Bhatt, R.S., Collier, K.J., Hoffman, B.J., Böhlen, P., Esch, F., Ling, N., Zeytin, F., Brazeau, P., Poonian, M.S. and Gage, L.P., *Proc. Natl. Acad. Sci. 80,* 14, 4311 (1983).

Herbert, E. and Uhler, M., *Cell 30,* 1 (1982).

Huttner, W.B., *Nature 299,* 273 (1982).

Ivell, R. and Richter, D., this volume.

Land, H., Grez, M., Ruppert, S., Schmale, H., Rehbein, M., Richter, D. and Schütz, G., *Nature 302,* 342 (1983).

Rapoport, T.A., Thiele, B.I., Prehn, S., Marbaix, G., Cleuter, Y., Hubert, E. and Huez, G., *Eur. J. Biochem. 87,* 229 (1978).

Richter, D., in "Progress in Nucleic Acid Research" (K. Moldave and W. Cohen, eds.), pp. 248. Academic Press, New York (1983).

Robbins, D.C., Blix, P.M., Rubenstein, A.H., Kanazawa, Y., Kosaka, K. and Tager, H.S., *Nature 291,* 679 (1981).

Sachs, H. and Takabatake, Y., *Endocrinology 75,* 943 (1975).

Schmale, H. and Richter, D., *FEBS Lett. 121,* 358 (1980).

Steiner, D.F., Docherty, K., Chan, S.J., Segundo, B.S. and Carroll, R., this volume.

PART V

Regulation and Metabolites

DEGRADATION OF NEUROPEPTIDES

Karl Bauer
Bernhard Horsthemke

Institut für Biochemie
Technische Universität Berlin
Berlin (West)

For neuropeptides, as for other biologically highly active substances, highly efficient inactivation mechanisms are likely to exist. It has been suggested that neuropeptides are degraded by neuropeptide-specific enzymes, which might even serve a regulatory function. So far, however, only general proteolytic enzymes have been identified in the brain and pituitary.

After enzyme fractionation and characterization it could be shown, for example, that the degradation of TRH is initiated by a pyroglutamate aminopeptidase and a post proline cleaving enzyme. LH-RH, Substance P and bradykinin are also hydrolyzed by various peptidases recognizing only certain structural elements of the substrates. The same is true for the enkephalins, which are degraded by aminopeptidases, a dipeptidyl aminopeptidase, a dipeptidyl carboxypeptidase ("angiotensin converting enzyme") and a thermolysin-like metallopeptidase.

The physiological functions of these enzymes remain to be elucidated. While some peptidases may be involved in removing neuropeptides from their target sites, other peptidases may serve a more general scavenger function at sites remote from the receptors. In no case would the clearance of neuropeptides represent a regulatory mechanism. The use of selective peptidase inhibitors and enzyme-specific antibodies should enable us to define the role of the individual neuropeptide-degrading enzymes more precisely.

Supported by the Deutsche Forschungsgemeinschaft

ISBN 0-12-417320-9

INTRODUCTION

An increasing number of neuropeptides have been identified in recent years, and most of them have been localized not only in selected areas of the Central Nervous System but also in specific cells of the gastrointestinal tract. These peptides act as neurohormones, neurotransmitters, neuromodulators or paracrine effector substances on a variety of target cells via binding to specific receptors. So far, however, the intracellular effector systems and the mechanisms of neuropeptide inactivation are poorly understood.

Since some neuropeptides are rapidly degraded *in vitro* by blood and tissue enzymes, it has been postulated that the biological inactivation of the neuropeptides is also catalyzed by these enzymes. Furthermore, it has been suggested that these enzymes even serve a dynamic control function by regulating the amount of neuropeptides available at the receptor level (Hooper, 1968; Griffiths et al., 1972; Knights et al., 1973; Griffiths et al., 1975; Prasad et al., 1976; Fridkin et al., 1977; Kuhl et al., 1978; Malfroy et al., 1978; Matsui et al., 1979; Advis et al., 1982). This hypothesis is based on the following assumptions:

1. Neuropeptides are degraded by neuropeptide-specific enzymes (e.g. oxytocin by "oxytocinase"; LH-RH by a LH-RH specific endopeptidase; enkephalins by "enkephalinase" etc.).
2. These neuropeptide-degrading enzymes exhibit a high affinity towards their substrates (Km-values in the nmolar range).
3. The increased biological activity of some analogues reflects their resistance against degradation rather than changes at the receptor level.
4. The activity of such neuropeptide-degrading enzymes is finally controlled by feedback-regulatory mechanisms (presumably due to the induction of enzyme synthesis either directly at the genomic level through estrogens, glucocorticoids, thyroid hormones or indirectly by some unknown mechanisms as in the case of morphine, which is not known to act directly at the genomic level).

If this hypothesis were true, we should also expect that

a. the activities of the neuropeptide-degrading and inactivating enzymes represent a limiting factor within the regulatory mechanisms, and that

b. changes in enzyme activities correlate in time with physiological, neuroendocrine or behavioral events.

To test this hypothesis it is prerequisitory to delineate the pathways of neuropeptide fragmentation and to evaluate the biochemical properties of the enzymes capable of hydrolyzing the neuropeptides.

DEGRADATION OF THYROTROPIN RELEASING HORMONE (TRH)

The rapid degradation of TRH by blood and tissue enzymes was already observed before the isolation of the tripeptideamide. It was also known that TRH is not degraded by general proteolytic enzymes such as aminopeptidases, carboxypeptidases, trypsin, chymotrypsin, pepsin, papain etc. (Burgus and Guillemin, 1970) and therefore it has been suggested that TRH might be degraded by a TRH-specific enzyme. After incubation of radiolabelled TRH with hypothalamic or hypophyseal tissue homogenates and chromatographic separation of the reaction mixture, Deamido-TRH, pyroGlu and histidyl-prolineamide could be identified as primary cleavage products (Prasad and Peterkofsky,1976; Bauer and Kleinkauf, 1980). Qualitatively the same fragmentation patterns were also obtained with homogenates of other tissues. These results already demonstrated that TRH is not degraded by only one enzyme but by two, ubiquitously distributed enzymes.

The hydrolysis of TRH at the pyroglutamyl-histidine bond is catalyzed by a pyroglutamate aminopeptidase (molecular weight ca. 28 000) which also hydrolyzes other pyroglutamyl-containing peptides such as LH-RH, neurotensin, gastrin as well as synthetic substrates (Armentrout and Doolittle, 1968; Szewczuk and Kwiatkowsky, 1970; Mudge and Fellows, 1973).

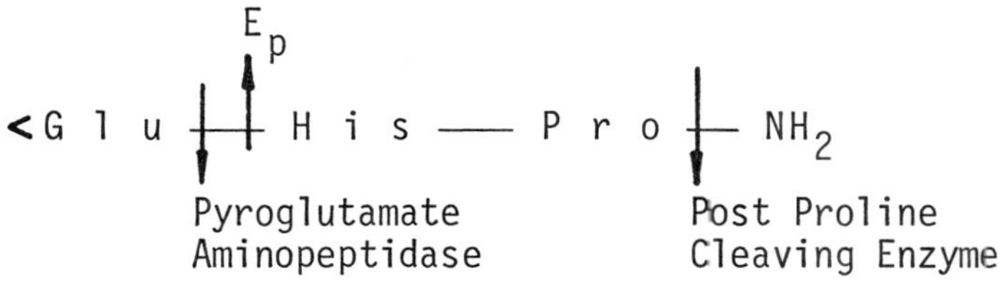

The deamidation of TRH is also not catalyzed by a specific "TRH-Deamidase", as suggested (Prasad and Peterkofsky, 1976), but by a post proline cleaving enzyme (Knisatschek and Bauer, 1979; Hersh and McKelvey, 1979; Andrews *et al.*, 1980; Koida and Walter, 1976), which hydrolyzes a variety of neuropeptides at internal proline-X peptide bonds.

Recently we could identify a particulate peptidase in the brain and pituitary (E_p), which also hydrolyzes TRH at the pyroGlu-His peptide bond. This enzyme seems to be identical to the peptidase described by Greaney *et al.*(1980), but it differs from the soluble pyroglutamate aminopeptidase in various physical and chemical properties.

DEGRADATION OF LUTEINIZING HORMONE-RELEASING HORMONE (LH-RH)

The characterization of the TRH-degrading enzymes already demonstrated that LH-RH can be hydrolyzed by the pyroglutamate aminopeptidase (E_1) and the post proline cleaving enzyme (E_2).

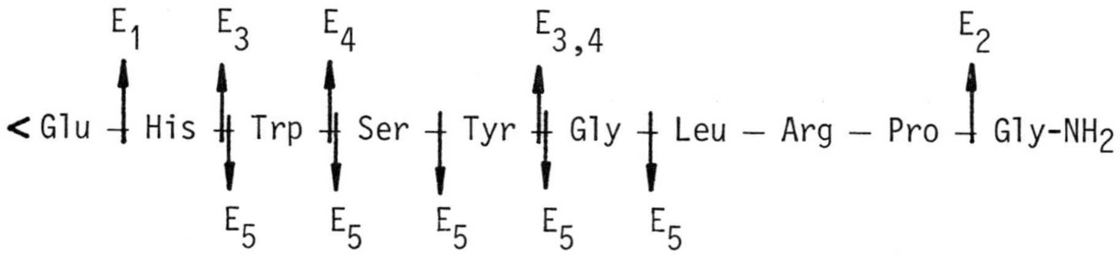

In addition, other enzymes have been identified, namely a non-chymotrypsin-like endopeptidase (E_3) (Horsthemke and Bauer, 1980; Horsthemke and Bauer, 1982), a cation-sensitive neutral endopeptidase (E_4) (Wilk and Orlowski, 1980; Horsthemke and Bauer, 1981) and a particulate LH-RH degrading enzyme (E_5) (Leblanc *et al*., 1980), which cleave the decapeptideamide at multiple *sites*. As in the case of TRH, a specific LH-RH degrading peptidase has not yet been found.

DEGRADATION OF SUBSTANCE P

A number of enzymes capable of degrading Substance P have been identified in brain and other tissues. Sequential removal of the Arg-Pro and Lys-Pro residues by the dipeptidyl aminopeptidase IV (E_1; post proline dipeptidyl aminopeptidase) (Heymann and Mentlein, 1978; Kato *et al*., 1980)leads to the C-terminal heptapeptideamide which *retains* full biological activity.

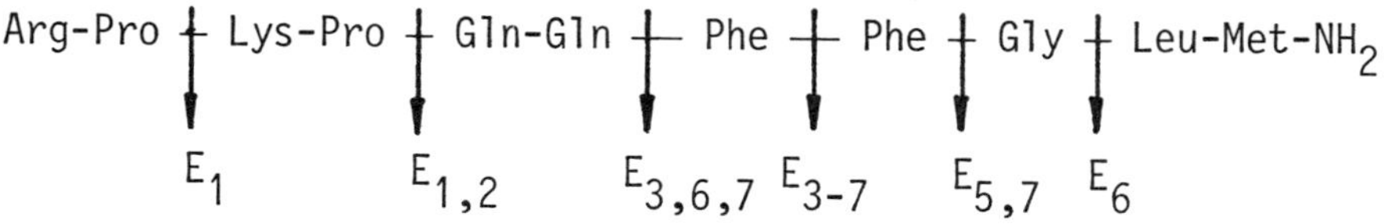

The same fragment is formed by the post proline cleaving enzyme (E_2) (Orlowski *et al*., 1979; Blumberg *et al*., 1980). Therefore, both *enzymes* could only contribute *to* the inactivation of Substance P if they act in concert with other enzymes such as aminopeptidases. Akopyan et al.(1979) described a neutral endopeptidase (E_3) which hydrolyzes Substance P probably between Gln-Phe and Phe-Phe. Purified lysosomal Cathepsin D (E_4) (Benuck et al., 1977) cleaves the undecapeptideamide between Phe-Phe. An enzyme called tonin (E_5) was reported to hydrolyze the Phe-Phe and Phe-Gly peptide bonds (Chrétien et al., 1980). A thermolysin-like metallopeptidase ("enkephalinase", E_6) cleaves at the amino side of the hydrophobic amino acids (Mumford et al., 1981; Matsas et al., 1983). Recently, a particle-bound Substance P degrading enzyme (E_7) has been isolated from human brain (Lee et al., 1981). The homogeneous enzyme preparation cleaves Substance P at multiple sites (Gln-Phe, Phe-Phe and Phe-Gly). This neutral endopeptidase of 40,000 - 50,000 molecular weight appears to exhibit a high specificity towards Substance P.

DEGRADATION OF BRADYKININ

Immunoreactive bradykinin has been localized to neuronal cells and fibers in rat brain (Correa et al., 1979; Hökfelt et al., 1980). Bradykinin can be degraded by a number of tissue and plasma exo- and endopeptidases, such as e.g. angiotensin converting enzyme (Erdös and Yang, 1970). Bradykinin degrading

Arg — Pro — Pro ↓ Gly — Phe ↓ Ser — Pro ↓ Phe — Arg

E_1 (Pro–Gly) $E_{2,3}$ (Phe–Ser) E_1 (Pro–Phe)

enzymes have also been detected in brain and pituitary. The cleavage of the peptide at the carboxyl side of proline residues is catalyzed by brain endopeptidase B (E_1) (Camargo et al., 1973; Oliveira et al., 1979; Carvalho and Camargo, 1981), an oligopeptidase which appears to be identical with the post proline cleaving enzyme (Wilk and Orlowski, 1979). The Phe-Ser bond is cleaved by the cation-sensitive neutral endopeptidase (Wilk and Orlowski, 1979; Wilk and Orlowski, 1980) and by brain endopeptidase A (Camargo et al., 1973; Oliveira et al., 1979; Carvalho and Camargo, 1981). All properties of the latter enzyme are similiar to those of the LH-RH degrading non-chymotrypsin-like endopeptidase (Horsthemke and Bauer, 1980; Horsthemke and Bauer, 1982; Camargo et al., 1982).

DEGRADATION OF ENKEPHALINS

The enkephalins are subject to extremely rapid degradation by various exopeptidases. This feature accounts for the short biological half-life of the pentapeptides.

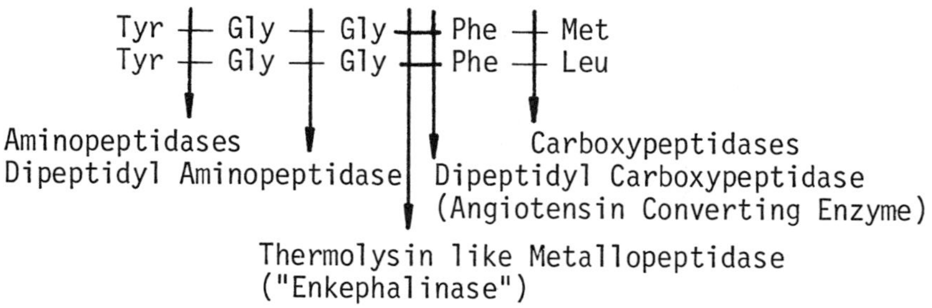

The removal of the aminoterminal tyrosine and thereby the inactivation of the pentapeptides is catalyzed by a variety of as yet not fully characterized aminopeptidases (Hambrook *et al.*, 1976; Lane *et al.*, 1977; Shaw and Cook, 1977; Hersh and McKelvey, 1981). In all systems studied so far, tyrosine was always found as the major metabolite of enkephalin degradation.

The observed fragmentation of the enkephalins at the Gly-Gly bond is apparently catalyzed by a so far not fully characterized dipeptidyl aminopeptidase (Gorenstein and Snyder, 1979). The relationship between this enzyme and the lysosomal dipeptidyl aminopeptidase extensively characterized by McDonald *et al.* (1968), remains to be established.

The cleavage of the Gly-Phe bond is catalyzed by the angiotensin converting enzyme (Erdös et al., 1978), a dipeptidyl carboxypeptidase of broad substrate specificity, and by an enzyme which has been termed "enkephalinase" since preliminary studies had indicated that this enzyme exhibits strict substrate specificity and high affinity towards enkephalins (Malfroy *et al.*, 1978). "Enkephalinase", however, appears to be identical or very similiar to a thermolysin-like neutral endopeptidase previously identified by Kerr and Kenney, 1974 (Almenoff *et al.*, 1981; Malfroy and Schwartz, 1982; Fulcher *et al.*, 1982).

EVALUATION OF THE BIOCHEMICAL STUDIES

The biochemical studies did not reveal any evidence for the presence of neuropeptide-specific enzymes in the brain or the pituitary. The primary specificity of the peptidases identified in these tissues is rather directed towards certain structural elements of the peptide (e.g. the aminoterminus or an amino acid side chain). This does not exclude the possibility that the rate of hydrolysis of a given peptide is influenced by neighboring groups or even the conformation of the substrate. The Km-values are in the micromolar range as expected for general proteolytic enzymes.

Such peptidases are capable of degrading an unlimited number of neuropeptides provided they contain the appropriate structural elements. It is clear, however, that these enzymes which lack the necessary selectivity and specificity of a regulation process, cannot fulfill a regulatory function themselves.

The characterization of these enzymes allowed for the synthesis of enzyme-specific substrates, inhibitors and degradation-resistant analogues suitable for further studies. Using enzyme-specific tests for the individual LH-RH degrading peptidases, we observed fluctuations in some peptidasic activities after physiological manipulation of experimental animals. These fluctuations, however, do not correlate in time with the neuroendocrine or behavioral events (Loudes et al., 1978; Bauer, et al., 1981) and most likely reflect changes within homeostatic adaption processes rather than specific changes within feedback regulatory mechanisms. This may also be true for the fluctuations in peptidasic activities observed in other cases. To date, there is no experimental evidence to support the view that neuropeptide metabolism is controlled by such enzymes.

After the elucidation of the substrate specificity and catalytic mechanism of the angiotensin converting enzyme, a potent inhibitor called captopril was synthesized (Ondetti et al., 1977). The effect of this inhibitor in whole animals is in agreement with the concept that in the periphery this enzyme is involved in controlling blood pressure. In the case of brain and pituitary enzymes, however, it is extremely difficult to interpret such pharmacological data. It is true that the administration of appropriate peptidase inhibitors prolongs and enhances the activity of brain peptides, especially if these are exogenously applied. This observation, however, does not necessarily imply that a given peptidase is directly involved in the peptidergic signal transmission, because the effective concentration of a neuropeptide at its target site can also be influenced by the hydrolysis of the diffusing peptides at sites remote from the receptor.Therefore, these data cannot be used to define precisely the biological function of these enzymes.

Interestingly, many neuropeptides which are resistant to hydrolysis by these peptidases (e.g. [D-Ser(tBu)^6]LH-RH(1-9) ethylamide, [D-Ala^2]Met-enkephalinamide and [<Glu^5-MePhe^8-Me-Gly^9]Substance P (5-11)) have prolonged and enhanced biological activity. As shown for several LH-RH agonists, however, there is no strict colinearity between the superactivity and resistance to degradation (Horsthemke et al., 1981). This indicates that increased receptor binding or stimulation must be considered also.

SUBCELLULAR LOCALIZATION OF NEUROPEPTIDE DEGRADING ENZYMES

Since a given peptidase can potentially act on several peptides and, vice versa, a given peptide could be degraded by various peptidases, the biochemical studies can not provide any direct information concerning the physiological function of these peptidases. The specificity of this function appears to be mainly governed by the localization of the peptidase in relation to a given peptide. Therefore it is most important to determine the distribution of these enzymes. In collaboration with Drs. P. Leblanc, C. Kordon (INSERM, Paris), S. Wattiaux-DeConinck and R. Wattiaux (Namur, Belgium) we determined the subcellular localization of several peptidases in rat pituitaries. These results are given in Table I.

The plasma membrane bound enzymes like the thermolysin-like metallopeptidase, the angiotensin converting enzyme, the particulate TRH-degrading peptidase and some aminopeptidases may potentially be involved in degrading extracellular neuropeptides, whereas intracellularly localized enzymes cannot act directly on extracellular neuropeptides and therefore, these enzymes are not directly involved in the peptidergic signal transmission.

CELLULAR LOCALIZATION

By screening primary cultures of neurons, astroblasts and astrocytes from murine brain for the presence of enkephalin-degrading enzymes we observed that bestatin-sensitive aminopeptidases represent the major enkephalin-degrading activity in all cases. The dipeptidyl aminopeptidasic activity is much higher in neuronal than in glial cultures, whereas the opposite is true for the "enkephalinase". Only trace amounts of

TABLE I

ENZYME	NEUROPEPTIDES KNOWN TO BE DEGRADED	LOCALIZATION
Angiotensin converting enzyme	*enkephalins, angiotensin I*	*plasma membrane*
Thermolysin-like metallopeptidase ("enkephalinase")	*enkephalins, Substance P*	*plasma membrane*
Partile-bound TRH-degrading peptidase	*TRH*	*plasma membrane*
Aminopeptidases	*enkephalins, oxytocin, vasopressin, angiotensin*	*ubiquitous, mainly plasma membranes*
Arylamidases	*enkephalins, oxytocin etc.*	*ubiquitous, mainly mitochondria*
Particle-bound LH-RH degrading enzyme	*LH-RH*	*mitochondria*
Particle-bound Substance P-degrading enzyme	*Substance P*	*mitochondria*
Post proline dipeptidyl aminopeptidase	*Substance P*	*ubiquitous*
Post proline cleaving enzyme	*LH-RH, Substance P, TRH etc.*	*cytosol*
Pyroglutamate aminopeptidase	*TRH, LH-RH, neurotensin*	*cytosol*
Non chymotrypsin-like endopeptidase	*LH-RH, bradykinin*	*cytosol*
Cation sensitive neutral endopeptidase	*LH-RH, Substance P, etc.*	*cytosol*

angiotensin converting enzyme have been found, which is known to be associated mainly with endothelial cells. Similiar observations concerning the enkephalin-degrading activities in cultures of rat astroblasts have been reported recently (Lentzen and Palenker, 1983).

These results demonstrate that certain plasma-membrane bound peptidases are not general, membrane-constituent enzymes, but are heterogeneously distributed between different cells in the brain, reflecting specialized functions. Although the structural and functional integrity of cell interaction in the brain is not conserved in primary cultures of neurons and glial cells, the present results suggest that in the case of the enkephalins bestatin-sensitive aminopeptidases on nerve cells could potentially be involved in the removal of enkephalins from the synaptic cleft, whereas the "enkephalinase" on glial cells may hydrolyze peptides which already left the synaptic cleft by diffusion. Ultimate proof, however, can only be obtained by further studies.

REGIONAL AND TISSUE DISTRIBUTION

Most of the neuropeptide-degrading enzymes described above have been found in various tissues. These findings suggest that a given peptidase fulfills different functions at different sites. For example, in the periphery angiotensin converting enzyme cleaves angiotensin II and thus fulfills an important function in the regulation of blood pressure. The physiological function of this enzyme in the brain remains to be elucidated. Within a given tissue, the peptidases are not necessarily distributed homogeneously. It could be shown for some enzymes that they are located only in specific brain regions or neuronal populations, suggesting again that such enzymes are concerned with specialized functions. However, in no case the distribution of the enzymes corresponds fully with any of the relevant peptidergic pathways known. Perhaps, we should not expect that such a strict correlation exists since we cannot expect that these neuropeptide-degrading enzymes are primarily concerned with the degradation of only one peptide. The information from these studies studies could eventually be used for a negative screening of neuropeptide-degrading enzymes. However, the question whether neuropeptide-degrading enzymes remove neuropeptides from the site of release and/or target interaction cannot be answered without histochemical methods, especially at the electron microscope level.

REFLECTIONS ABOUT THE PHYSIOLOGICAL FUNCTION OF NEUROPEPTIDE-DEGRADING ENZYMES

DEGRADATION OF NEUROPEPTIDES BEFORE SECRETION

For the neuropeptide-synthesizing cells, some as yet unknown mechanisms must exist to regulate the intracellular concentration of the neuropeptides. For example, under conditions when secretion is suppressed, these mechanisms might act as a security device system. Since the sites of synthesis, storage and release are necessarily separated from the site of degradation, the neuropeptide-degrading enzymes cannot act directly on these peptides and therefore, these enzymes cannot fulfill a regulatory function. The mechanisms which determine the translocation of the peptides (such as e.g. the mechanisms leading to the crinophagic disposal of secretion granules etc.) apparently represent the limiting and thus regulatory control elements which finally determine the biological inactivation of the neuropeptides. These mechanisms are yet unknown and experimentally most difficult to evaluate since hydrolysis of the neuropeptides does not lead to specific metabolites but only to cell constituent amino acids. The availability of enzyme-specific inhibitors might provide the tools to explore these mechanisms and also to obtain some information concerning, for example, the turnover rates of these substances.

DEGRADATION OF NEUROPEPTIDES DURING TRANSPORT TO THEIR TARGET SITES

Enzymes present in the medium of intercellular communication may potentially modulate the availability of the peptides at their target sites. In this case, the specificity of the degrading process can not be due to the specific localization of the enzyme but must be due to the specificity of the enzyme itself. Therefore, we should expect that an enzyme with such a physiological function should exhibit a high degree of substrate specificity. Moreover, the activity of such an enzyme should be finally controlled by feedback-regulatory mechanisms.

The TRH-degrading serum enzyme may potentially fulfill such a regulatory function. Like the pyroglutamate aminopeptidase, the TRH-degrading serum enzyme catalyzes the hydrolysis of TRH at the pyroglutamyl-histidine bond. However, the TRH-degrading serum enzyme exhibits chemical and physical characteristics distinctly different from those of the pyroglutamate aminopeptidase (Taylor and Dixon, 1978; Bauer and Nowak, 1979). Moreover and most importantly, the TRH-degrading serum enzyme exhibits a high degree of substrate specificty (Bauer and Nowak, 1981). Unlike the pyroglutamate aminopeptidase this enzyme does not degrade LH-RH, neurotensin and other pyroglutamyl-containing peptides. Together with the observations that in the rat the activity of the TRH-degrading serum enzyme is controlled by thyroid hormones (Bauer, 1976; White et al., 1976) and drastically alters with developmental changes (Oliver et al.,1974; Neary et al., 1976), these results suggest that the TRH-degrading serum enzyme may represent a functional control element within the mechanisms regulating the availability of the neuropeptide to the trophic cells of the pituitary.

It should be noted, however, that for other neuropeptides a specific enzyme has not been identified so far. Most neuropeptide-degrading enzyme present in serum or CSF are general peptidases which may serve general scavenger functions.

DEGRADATION OF NEUROPEPTIDES AFTER BINDING TO THEIR RECEPTORS

A. By Intracellular Enzymes After Reuptake Into the Secreting Cell

Catecholamines are known to be removed from the synaptic cleft by reuptake into the secreting cell, where they are oxidized by the mitochondrial monoamine oxidase or methylated by catecholamine-O-methyl-transferase. By analogy with these substances, peptidergic transmitters may also be taken up by the secreting cell and hydrolyzed by intracellular enzymes. So far, however, there is little evidence for such a process. Recently, the uptake of enkephalins by synaptosomal vesicle preparations has been reported (Takeda et al., 1982; Roy et al., 1982). Further studies are necessary to explore whether such mechanisms contribute to the inactivation of neuropeptides in general.

B. By Intracellular Enzymes After Uptake Into the Target Cell

As shown for several proteohormones and some neuropeptides, peptides may enter the target cell by receptor mediated endocytosis. The fate of the peptide-receptor complex is not known in most cases, but conceivably the peptides are degraded by lysosomal or cytosolic enzymes. The internalization of peptides is a relatively slow process, but may represent an adequate mechanism for the removal of neuropeptides exhibiting an extremely low K_D.

C. By Enzymes In the Vicinity of Neuropeptide Receptors

By analogy with acetylcholinesterase it is easy to conceive that neuropeptides may be removed from their site of action by hydrolysis. This may especially occur in such cases of peptidergic transmission, where the synaptic cleft has to be cleared rapidly for the transmission of the next signal. Only in this case would the removal of the neuropeptides be due to the action of peptidases. Not necessarily, such enzymes would have to have a high affinity and specificity for a given peptide. However, we should expect that such enzymes exhibit a high turnover number and that they are present at these sites in sufficiently high concentrations.

THE ROLE OF NEUROPEPTIDE-DEGRADING ENZYMES IN TURNING OFF PEPTIDERGIC SIGNALS

In the course of turning off peptidergic signals, three stages may generally be distinguished from each other:

1. *Termination of action.* The action of a neuropeptide seems to be terminated exclusively at the receptor level, either by dissociation of the receptor-peptide complex or by transition to a desensitized form. There is no evidence for the hydrolysis of a ligand while binding to its receptor.

2. *Clearance of the site of target interaction.* To clear the target site for the transmission of the next signal, all peptide molecules have to be removed. This may occur by diffusion, cellular uptake or hydrolysis.

3. *Inactivation of neuropeptides.* The peptides are degraded either by enzymes located in the vicinity of the receptors or at sites remote from the receptors.

It is clear, that neuropeptide-degrading enzymes themselves are not the "switch" for turning off peptidergic signals. They can only be involved in the late steps of this process, thus fulfilling a scavenger function rather than a direct regulatory role. This does not exclude the possibility that their concentration is subject to homeostatic adaption processes. Obviously, neuropeptide-degrading enzymes fulfill a biologically most important function for the organism: they maintain the system of cellular communication by clearing the site of target interaction , preventing the accumulation of neuropeptides and thus the poisoning of the system.

REFERENCES

1. Advis, J.P., Krause, J.E., and McKelvey, J.F., *Endocrinology 110*, 1238 (1982).
2. Akopyan, T.N., Arutunyan, A.A., Organisyan, A.I., Lajtha, A., and Galoyan, A.A., *J. Neurochem. 32*, 629 (1979).
3. Almenoff, J., Wilk, S., and Orlowski, M., *Biochem. Biophys. Res. Commun. 102*, 206 (1981).
4. Andrews, P.C., Hines, C.M., and Dixon, J.E., *Biochemistry 20*, 7082 (1980).
5. Armentrout, R.M., and Doolittle, R.F., *Arch. Biochem. Biophys. 132*, 89 (1968).
6. Bauer, K., *Nature 259*, 591 (1976).
7. Bauer, K., and Nowak, P., *Eur. J. Biochem. 99*, 239 (1979).
8. Bauer, K., and Kleinkauf, H., *Eur. J. Biochem. 106*, 107 (1980).
9. Bauer, K., and Nowak, P., *Eur. J. Biochem. 118*,173 (1981).
10. Bauer, K., Beier, S., Horsthemke, B., Knisatschek, H., and Sievers, J., *Exptl. Brain Res. Suppl. 3*, 93 (1981).
11. Benuck, M., Grynbaum, A., and Marks, N., *Brain Res. 143*, 181 (1977).
12. Blumberg, S., Teichberg, V.I., Charli, J.L., Hersh, L.B., and McKelvey, J.F., *Brain Res. 192*, 447 (1980).
13. Burgus, R., and Guillemin, R., *in* "Hypophysiotropic Hormones of the Hypothalamus" (J. Meites ed.), p. 227. Williams and Wilkins, Baltimore (1970).
14. Camargo, A.C.M., Shapanka, R., and Greene, L.J., *Biochemistry 12*, 1838 (1973).
15. Carvalho, K.M., and Camargo, A.C.M., *Biochemistry 20*, 7082 (1981).
16. Camargo, A.C.M., Fonesca, M.J.V., Caldo, H., and Carvalho, K.M., *J. Biol. Chem. 257*, 9265 (1982).
17. Chrétien, M., Lee, C.M., Sandberg, B.E.B., Iversen, L.L., Boucher, R., Seidah, N.G., and Genet, J., *FEBS Lett. 113*, 173 (1980).
18. Correa, F.M.A., Innis, R.B., Uhl, G.R., and Snyder, S.H., *Proc. Nat. Acad. Sci. 76*, 1489 (1979).
19. Erdös, E.G., and Yang, H.Y.T., *in* "Handbook of Experimental Pharmacology" (E.G. Erdös, ed.), p.289. Springer Verlag, Berlin (1970).
20. Erdös, E.G., Johnson, A.R., and Boyden, N.T., *Biochem. Pharmacol. 27*, 843 (1978).
21. Fridkin, M., Hazum, E., Baram, T., Lindner, H.R., and Koch, Y., *in* "Proceedings of the Vth American Peptide Symposium" (M. Goodman and J. Meienhofer, eds.), p.193. Halsted Press, New York, (1977).

22. Fulcher, I.S., Matsas, R., Turner, A.J., and Kenney, A.J., *Biochem. J. 203*, 519 (1982).
23. Gorenstein, C., and Snyder, S.H., *Life Sci. 25*, 2065 (1979).
24. Greaney, A., Phelan, J., and O'Cuinn, G., *Biochem. Soc. Trans. 8*, 423 (1980).
25. Griffiths, E.C., and Hooper, K.C., *Acta endocr.(Kbh.) 69*, 249 (1972).
26. Griffiths, E.C., Hooper, K.C., Jeffcoate, S.L., and Holland D.T., *Brain Res. 88*, 384 (1975).
27. Hambrook, J.M., Morgan, B.A., Rance, M.J., and Smith, C.F. C., Nature *262*, 782 (1976).
28. Hersh, L.B., and McKelvey, J.F., *Brain Res., 168*, 553 (1979).
29. Hersh, L.B., and McKelvey, J.F., *J. Neurochem., 13*, 171 (1981).
30. Heymann, E., and Mentlein, R., *FEBS Lett. 91*, 360 (1978).
31. Hökfelt, T., Johansson, O., Ljungdahl, A., Lundberg, J.M., and Schulzberg, M., Nature *284*, 515 (1980).
32. Hooper, K.C., *Biochem. J. 110*, 151 (1968).
33. Horsthemke, B., and Bauer, K., Biochemistry *19*, 2867 (1980).
34. Horsthemke, B., Knisatschek, H., Rivier, J., Sandow, J., and Bauer, K., *Biochem, Biophys. Res. Commun. 100*, 753 (1981).
35. Horsthemke, B., and Bauer, K., Biochemistry *21*, 1033 (1982).
36. Kato, T., Hama, T., and Nagatsu, T., *J. Neurochem. 34*, 602 (1980).
37. Kerr, M.A., and Kenney, A.J., *Biochem. J. 137*, 477 (1974).
38. Knights, E.B., Baylin, S.B., and Foster, G.V., Lancet *II*, 719 (1973).
39. Knisatschek, H., and Bauer, K., *J. Biol. Chem., 254*, 10936 (1979).
40. Koida, M., and Walter, R., *J. Biol. Chem. 251*, 7593 (1976).
41. Krause, J.E., Advis, J.P., and McKelvey, J.F., *Biochem. Biophys. Res. Commun. 108*, 1475 (1982).
42. Kuhl, H., Rosniatowski, C., and Taubert, H.-D., *Acta endocr. (Kbh.) 87*, 476 (1978).
43. Lane, A.C., Rance, M.J., and Walter, D.S., Nature *269*, 75 (1977).
44. Leblanc, P., Pattou, E., L'Hérithier, A., and Kordon, C., *Biochem. Biophys. Res. Commun. 96*, 1457 (1980).
45. Lee, C.M., Sandberg, B.E.B., Hanley, M.R., and Iversen, L. L., *Eur. J. Biochem. 114*, 314 (1981).
46. Lentzen, H., and Palenker, J., *FEBS Lett. 153*, 93 (1983).
47. Loudes, C., Josepho-Bravo, P., Leblanc, P., and Kordon, C., *Biochem. Biophys. Res. Commun. 83*, 921 (1978).
48. Malfroy, B., Swerts, J.P., Guyon, A., Roques, B.P., and Schwartz, J.C., Nature *276*, 523 (1978).
49. Malfroy, B., and Schwartz, J.C., *Biochem. Biophys. Res. Commun. 106*, 276 (1982).

50. Matsas, R., Fulcher, I.S., Kenney, A.J., and Turner, A.J., *Proc. Nat. Acad. Sci. 80*, 3111 (1983).
51. Matsui, R., Prasad, Ch., and Peterkofsky, A., *J. Biol. Chem. 254*, 2439 (1979).
52. McDonald, J.K., Reilly, T.J., Zeitmann, B.B., and Ellis, S., *J. Biol. Chem. 243*, 2028 (1968).
53. Mudge, A.W., and Fellows, R.E., Endocrinology *93*,1428 (1973).
54. Mumford, R.A., Pierzchala, P.A., Strauss, A.W., and Zimmermann, M., *Proc. Nat. Acad. Sci. 78*, 6623 (1981).
55. Neary, J.T., Kieffer, J.C., Federico, P., Mover, H., Maloof, F., and Soodak, M., Science *193*, 7593 (1976).
56. Oliver, C., Taurog, A., and Porter, J.C., *Nouv. Press Med. 3*, 1941 (1974).
57. Oliveira, M.A., Martins, A.R., and Camargo, A.C.M., Biochemistry *15*, 1967 (1969).
58. Ondetti, M.A., Rubin, B., and Cushman, D.W., Science *196*, 441 (1977).
59. Orlowski, M., Wilk, E., Pearce, S., and Wilk, S., *J. Neurochem. 33*, 461 (1979).
60. Prasad, Ch., and Peterkofsky, A., *J. Biol. Chem. 251*, 3229 (1976).
61. Roy, B.P., Jamal, I., and Go, I., *Life Sci. 31*, 2307 (1982)
62. Shaw, S.G., and Cook, W.T., Nature *274*, 816 (1977).
63. Szewczuk, A., and Kwiatkowsky, J., *Eur. J. Biochem. 15*, 92 (1970).
64. Takeda, M., Takeda, F., Matsumoto, F., Tanaka, R., and Konno, K., *Brain Res. 234*, 319 (1982).
65. Taylor, W.L., and Dixon, J.E., *J. Biol. Chem. 253*, 6934 (1978).
66. White, N., Jeffcoate, S.L., Griffiths, E.C., and Hooper, K.C., *J. Endocrinol. 71*, 1 (1976).
67. Wilk, S., and Orlowski, M., *Biochem. Biophys. Res. Commun. 90*, 1 (1979).
68. Wilk, S., and Orlowski, M., *J. Neurochem. 35*, 1172 (1980).

METABOLITES OF ARGININE-VASOPRESSIN AND OXYTOCIN ARE HIGHLY POTENT NEUROPEPTIDES IN THE BRAIN

J. Peter H. Burbach
Gabor L. Kovács[1]
Xin-Chang Wang
David de Wied

Rudolf Magnus Institute for Pharmacology
Medical Faculty, University of Utrecht
Utrecht, The Netherlands

I. INTRODUCTION

The nonapeptides arginine-vasopressin (AVP-(1-9), AVP) and oxytocin (OXT-(1-9), OXT) are regulatory peptides for peripheral as well as central functions. The peptides are widely distributed over the central nervous system via axonal fiber systems originating in the hypothalamus (Buijs, 1978; Weindl and Sofroniew, 1980). AVP and OXT have central activities which are apparent in modulation of memory processes (De Wied, 1980), control of body temperature (Cooper et al., 1979) and brain development (Boer et al., 1980), induction of maternal behavior (Pedersen and Prange, 1979), sexual behavior (Bohus, 1977; Södersten et al., 1983) and development of tolerance and addiction to narcotic drugs (Van Ree and De Wied, 1981). The memory effects of these nonapeptides are evident in behavioral situations (De Wied, 1980) and have been found in several memory disturbed patients (Jolles, 1983; Ferrier et al., 1980). Characteristic for AVP is the memory improving influence; OXT at the other hand has amnesic activity.

Unlike the peripheral activities, the central activities of AVP and OXT are not confined to the nonapeptide structure

[1]*Present address: Department of Pathophysiology, University Medical School, Szeged, Hungary.*

ISBN 0-12-417320-9

itself. For instance, the des-GlyNH$_2$ derivative of AVP (DGAVP, AVP-(1-8)) has virtually lost its antidiuretic and vasopressor activity, but is fully potent in behavioral tests (De Wied et al., 1972). In addition, in structure-activity studies synthetic peptide fragments of AVP and OXT appeared to possess behavioral activities different or independent of those of the parent peptides indicating multiple sites of biological activity in the nonapeptide structure (De Wied, 1976; 1980; De Wied and Bohus, 1978). Based on these observations it has been put forward that AVP and OXT could serve as precursors for oligopeptides with more potent or selective central activities (De Wied, 1976; 1980; De Wied and Bohus, 1978). We have approached the identification of such neuropeptides by characterization of the products of proteolytic conversion of AVP and OXT by brain peptidases (Burbach et al., 1980a,b; Burbach and De Wied, 1981). This chapter describes the identification, mechanisms of formation, and endogenous presence of major metabolites of AVP and OXT in brain. These metabolites exclusively have central activity and are exceedingly more potent in affecting memory consolidation than AVP and OXT. It is indicated that these peptides may represent a novel class of neuropeptides with selective central effects.

II. IDENTIFICATION OF AVP- AND OXT METABOLITES

In order to obtain metabolites of AVP and OXT in quantities allowing isolation and subsequent chemical and biological characterization, peptide fragments were generated by exposing AVP and OXT to proteolytic activities present in a brain synaptic membrane preparation. Peptides accumulating during proteolytic digestion were isolated by HPLC. After prolonged incubation of AVP a major metabolite (retention time 15 min, fig. 1A) was obtained. It represented a single peptide

---

*FIGURE 1. Reversed-phase HPLC of metabolites of AVP (A) and OXT (B) generated by brain synaptic membranes. AVP (7.5×10^{-5} M) and OXT (5×10^{-5} M) were incubated with membranes (2.5 mg protein/ml) in sodium phosphate, pH 7.0, at 37° C for 6 h. Membrane free media were fractionated by HPLC on a μBondapak C18 column using 10 mM ammonium acetate, pH 4.15, (solvent A) and 0.15% acetic acid in methanol (solvent B) at a flow rate of 2 ml/min. For AVP digests a concave gradient from 0 to 40% B in 30 min was used, for OXT digests a linear gradient from 10 to 40% B in 20 min. The arrows indicate the elution position of identified peptides. Components marked * were derived from the membrane fraction.*

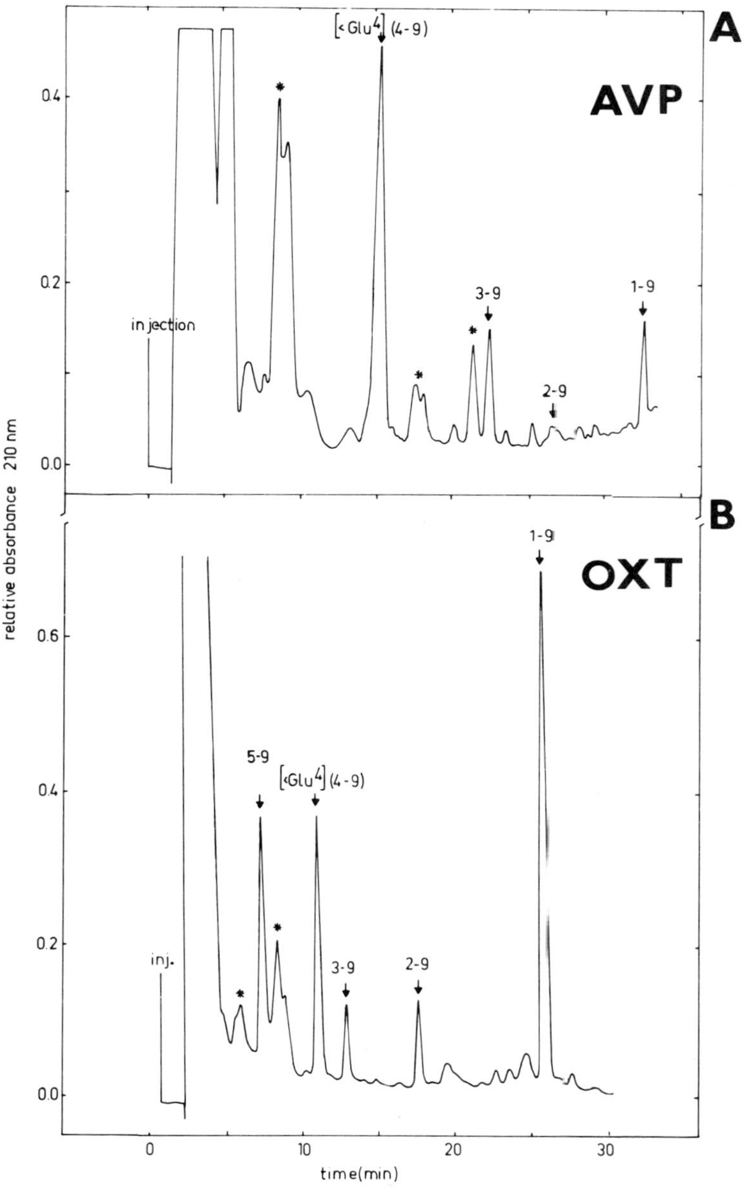
A
AVP
[<Glu4] (4-9)
3-9
1-9
2-9
injection
B
OXT
1-9
5-9
[<Glu4] (4-9)
3-9
2-9
inj.
relative absorbance 210 nm
0.0
0.2
0.4
0.6
0
10
20
30
time(min)

with amino acid composition Cys(O_3H)$_{2.0}$, Asp$_{1.0}$, Glu$_{0.9}$, Arg$_{1.0}$ and Gly$_{1.0}$ (Pro is not detected by the analysis technique employing o-phthaldialdehyde) and showing dansyl-Cys-OH H-Cys-OH in the end group analysis using dansyl chloride. Dansyl-Asp was detected after treatment of the peptide with pyroglutamate aminopeptidase (Burbach et al., 1983a). These analyses showed that the peptide was <Glu-Asn-Cys(Cys)-Pro-Arg-Gly-NH_2,[<Glu^4,Cyt^6]AVP-(4-9), the C-terminal hexapeptide of AVP having an intact disulfide bridge linking Cys^1 to the peptide backbone and a N-terminal pyroGlu residue. The structure is indicated in figure 4.

Incubation of OXT under comparable conditions resulted in accumulation of a major product eluting at 10.2 min in HPLC (fig. 1B). Chemical analyses demonstrated that this peptide was <Glu-Asn-Cys(Cys)-Pro-Leu-Gly-NH_2 ([<Glu^4,Cyt^6]OXT-(4-9)) (Burbach et al., 1983b).

In addition to these two metabolites a more complete set of related peptide fragments was isolated and identified from digests of AVP and OXT incubated at lower substrate concentration and for shorter time (Burbach and Lebouille, 1983). These conditions yielded considerable amounts of [Cyt^6]AVP-(2-9) and [Cyt^6]AVP-(3-9). In addition to [<Glu^4,Cyt^6]OXT-(4-9), the metabolites [Cyt^6]OXT-(2-9), [Cyt^6]OXT-(3-9), [Cyt^6]OXT-(4-9), and [Cyt^6]OXT-(5-9) were characterized. These peptides are also found under the conditions of fig. 1A and B as indicated. For structures see figure 4.

III. BIOLOGICAL ACTIVITIES OF AVP- AND OXT METABOLITES

The identified peptides [<Glu^4,Cyt^6]AVP-(4-9) and [<Glu^4,Cyt^6]OXT-(4-9) were chemically synthesized to study their biological activities (Van Nispen et al., 1983). Since AVP-(1-9) does not require its C-terminal Gly-NH_2 residue for its central activities (De Wied et al. 1972), the des-Gly-NH_2 derivatives of these peptides were synthesized and investigated as well.

A. Central Activities

The activity of metabolites in affecting brain functions, in particular those related to memory processes, was investigated in a passive avoidance situation (Ader and De Wied, 1972; fig. 2) by intracerebroventricular administration.

The AVP fragment [<Glu^4,Cyt^6]AVP-(4-9) was active in facilitating passive avoidance behavior in extremely low doses

of 0.03 and 0.10 pg, while the minimal effective dose of the parent peptide AVP-(1-9) was 100 pg (fig. 2A). The U-shaped dose-response relationship found with [<Glu^4,Cyt^6]AVP-(4-9) was not seen with the des-glycinamide derivative [<Glu^4,Cyt^6]-AVP-(4-8). This peptide was active over the tested dose range of 0.10 to 10 pg (fig. 2A). Thus, these AVP fragments are at least a thousand times more potent than AVP itself in this behavioral test (Burbach et al., 1983a).

Similarly to OXT, the OXT fragment [<Glu^4,Cyt^6]OXT-(4-9) attenuated the passive avoidance response, but it was about a hundred times more potent (Burbach et al., 1983b). The minimal effective dose of OXT-(1-9) was 100 pg, while for [<Glu^4,Cyt^6] OXT-(4-9) it was only 1 pg (fig. 2B). Again a U-shaped dose-response relationship was observed, which was absent in the (4-8)-derivative. However, [<Glu^4,Cyt^6]OXT-(4-8) was less potent but had a more stable behavioral effect than the (4-9)-peptide. The C-terminal pentapeptide of OXT, [Cyt^6]OXT-(5-9) was devoid of effect on memory consolidation in the tested doses (fig. 2B), indicating that the <Glu^4 residue is essential for this biological activity.

B. *Peripheral Activity*

The AVP fragments [<Glu^4,Cyt^6]AVP-(4-9) and -(4-8) did not show any vasopressor activity (Burbach et al., 1983a). In an *in vivo* bioassay (Dekanski, 1952) intravenous injection of 10 ng AVP caused a significant rise in blood pressure; [<Glu^4,Cyt^6]AVP-(4-9) and -(4-8) in doses of 10 to 500 ng were inactive and failed to inhibit the effect of AVP.

Peripheral agonistic and antagonistic effects of [<Glu^4, Cyt^6]OXT-(4-9) were tested on the isolated uterus *in vitro* of estradiol treated rats. The lowest concentration of OXT for maximal contraction was 0.14 ng/ml. Concentrations of [<Glu^4,Cyt^6]OXT-(4-9) ranging from 0.001 to 100 ng/ml did not show any uterotonic effect. The uterotonic effect of OXT was not inhibited by a thousand-fold molar excess of the OXT metabolites (Burbach et al., 1983b).

These data demonstrated that the hexapeptide metabolites lacked the classical endocrine activities of AVP and OXT while they have an outspoken central effect, suggesting a selective function of these peptides in the brain.

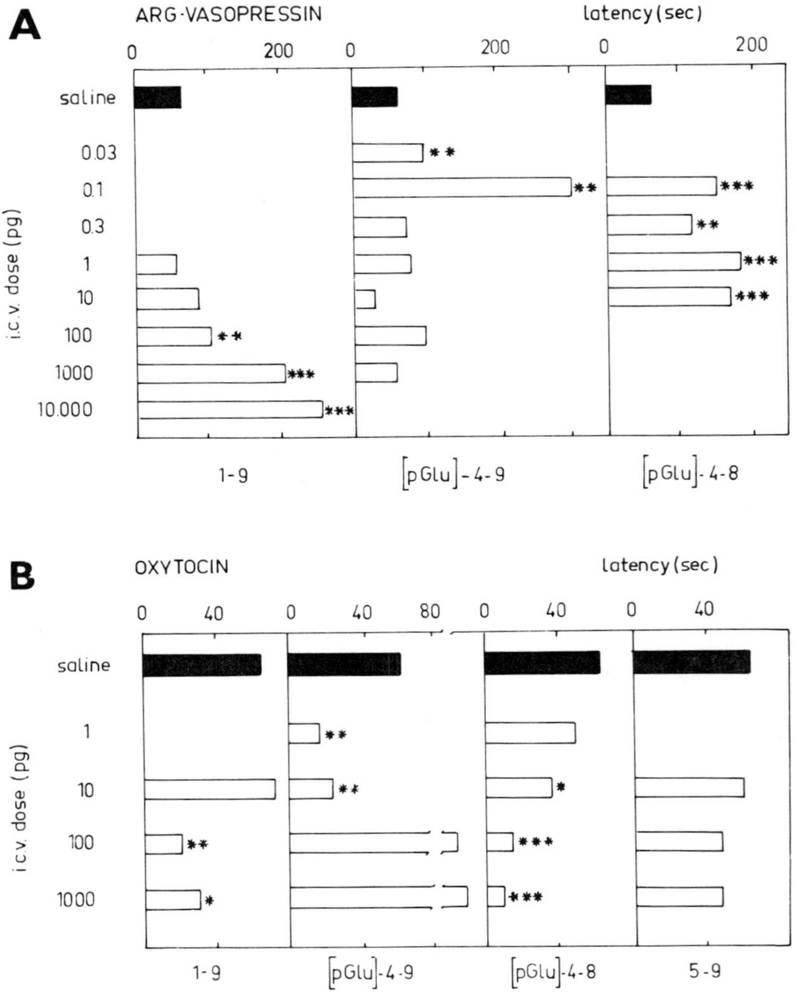

FIGURE 2. Activity of metabolites of AVP (A) and OXT (B) on memory consolidation as tested in a passive avoidance situation.

Passive avoidance behavior was studied in a one-trial learning, step-through situation according to Ader and De Wied (1972). Animals were placed on an illuminated platform attached to a dark compartment and were allowed to habituate to the situation. At the end of the adaptation sessions the animals received an unescapable electric footshock (0.25 mA during 2 sec) upon entering the dark compartment. Subsequently peptides or saline were administered by intracerebroventricular injection through a cannula placed in the right lateral ventricle. The influence of substances on the storage

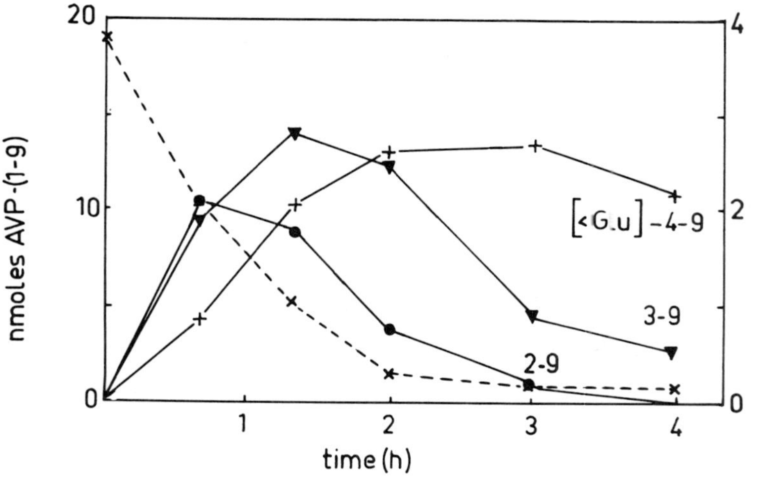

FIGURE 3. *Time course of formation of AVP metabolites. [^{14}C-Gly-NH$_2$^{9}]AVP-(1-9) was incubated with synaptic membranes and at time intervals aliquots were taken for HPLC analysis. The disappearance of AVP-(1-9) (x) and the accumulation of [Cyt^6]AVP-(2-9) (●), -(3-9) (▼), and [<Glu^4,Cyt^6]AVP-(4-9) (+) is depicted.*

---

of this aversive experience (memory consolidation) was assessed by measuring the latency to re-enter the dark compartment after placing the animals on the illuminated platform 24 h later.

*Statistical significance (Mann-Whitney U-test): *$p < 0.05$, **$p < 0.01$, ***$p < 0.001$, as compared to saline.*

IV. MECHANISM OF FORMATION OF ACTIVE AVP- AND OXT METABOLITES AND SOME PROPERTIES OF THE PEPTIDASE ACTIVITY

The nature of identified metabolites of AVP and OXT having truncated N-termini suggested that they were formed by consecutive aminopeptidase cleavage. Direct evidence for the involvement of aminopeptidase activity was obtained from time course studies using [^{14}C-GlyNH$_2$^{9}]- and [^{14}C-Tyr^2]-labeled AVP and OXT (Burbach and Lebouille, 1983). Proteolysis of AVP-(1-9) resulted in initial formation of [Cyt^6]AVP-(2-9), rapidly followed by [Cyt^6]AVP-(3-9) (fig. 3). The subsequent decline of these products and the accumulation of [<Glu^4,Cyt^6]-AVP-(4-9) (fig. 3) demonstrated the precursor-product interrelationships of these AVP metabolites and corroborated the aminopeptidase cleavage pattern. The eventual accumulation of [<Glu^4,Cyt^6]AVP-(4-9) is promoted by the formation of the pyroGlu residue which protects the peptide against further aminopeptidase cleavage. The production of C-terminal AVP- and OXT fragments involves an identical aminopeptidase cleavage pattern, but differs in rate of enzymatic cleavage; the half-life of AVP *in vitro* is approximately five times shorter than that of OXT (Burbach and Lebouille, 1983; Burbach et al., 1982). The related nonapeptide Arg-vasotocin (AVT) is processed in a similar fashion by rat as well as chicken brain membranes (Wang et al., 1983a). Therefore, the proteolytic processing by aminopeptidase activity represents a common metabolic pathway for AVP, OXT and AVT (fig. 4).

In earlier studies some properties of aminopeptidase activity acting on OXT-(1-9) were obtained (Burbach et al., 1980). This aminopeptidase activity, as measured by the release of ^{14}C-Tyr from [^{14}C-Tyr^2]OXT-(1-9), was most concentrated in synaptosomal and microsomal preparations of brain tissue and was relatively low in cytosol and mitochondrial fractions. It had a narrow pH optimum of pH 7.0. Synaptic membrane preparations of various brain areas differed slightly in activity of this peptidase: highest activities of the peptidase was found in the medial basal hypothalamus, the gyrus dentatus of the hippocampus and the dorsal raphe area (Burbach et al., 1980). Although activity was found in all brain areas investigated, a differential distribution of peptidase activity may indicate that metabolism of AVP and OXT can differ locally with possible consequences for the levels of active metabolites.

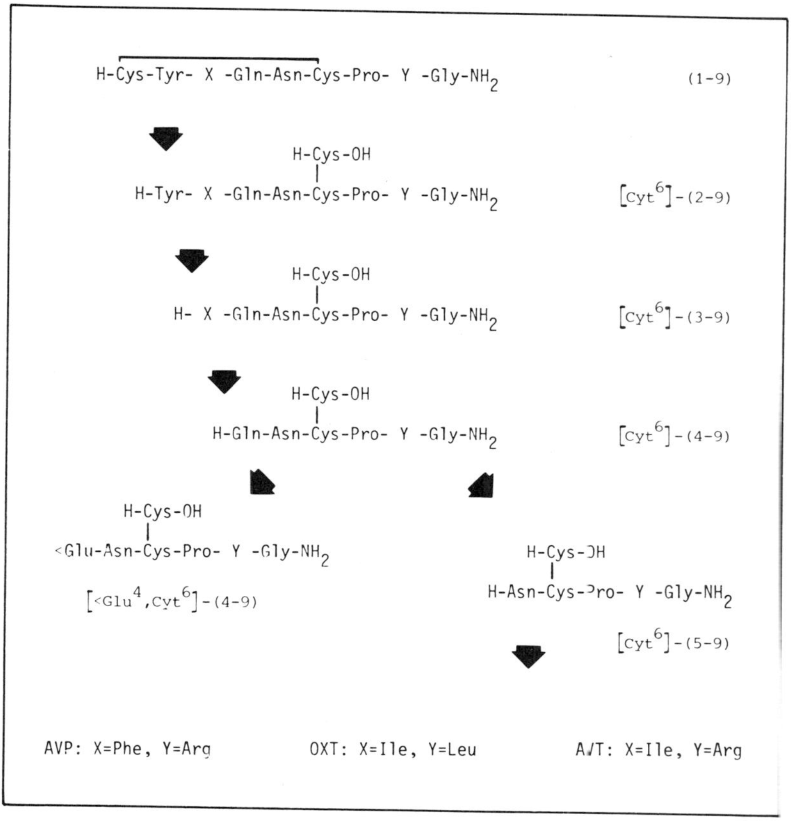

FIGURE 4. Pathway of proteolytic conversion of AVP, OXT, and AVT by brain synaptic membranes and structure of metabolites.

The peptides are those identified in digests of AVP, OXT, and AVT (Burbach and Lebouille, 1983; Burbach et al., 1980b, 1983a, b; Wang et al., 1983a). The sequence of proteolytic events has been derived from time course studies (Burbach and Lebouille, 1983).

V. ENDOGENOUS PRESENCE OF AVP METABOLITES

The findings of *in vitro* generated metabolites of AVP and OXT with potent activities in the brain urged to investigate the endogenous presence of these peptides in brain tissue. A radioimmunoassay system was developed that recognized AVP-(1-9), [Cyt^6]AVP-(2-9), -(3-9) and -(5-9), and [<Glu^4, Cyt^6]AVP-(4-9) (Wang et al., 1983b); these AVP metabolites cross-reacted between 50% ([Cyt^6]AVP-(2-9)) and 13% ([Cyt^6]-AVP-(5-9)) as compared to AVP-(1-9). Using this radioimmunoassay system in combination with HPLC fractionation of brain and pituitary extracts intact AVP-(1-9) and immunoreactive components comigrating with [Cyt^6]AVP-(2-9), [Cyt^6]AVP-(3-9), [<Glu^4,Cyt^6]AVP-(4-9) and [Cyt^6]AVP-(5-9) were detected. In addition, evidence for the presence of peptides with reduced disulfide bridge, in particular [<Glu^4]AVP-(4-9) and AVP-(5-9) was obtained. Control experiments using ^{3}H-AVP-(1-9) excluded that the AVP fragments were generated during extraction. It is therefore concluded that they represent endogenous peptides in brain tissue (Wang et al., 1983b).

The levels of AVP metabolites were markedly lower than that of intact AVP-(1-9). In brain the levels ranged between 1 to 20% of the AVP-(1-9) content. These peptides were found in several brain areas including hypothalamus, amygdala, septum, hippocampus and medulla oblongata with regional variations. Highest amounts of metabolites (approximately 4 to 20% of the AVP content) were found in amygdala, septum and hippocampus. Strikingly no significant amounts of AVP metabolites were detected in the pituitary gland. Only a trace of AVP metabolites, being about 0.05% as compared to AVP-(1-9) was found. This finding seems to indicate that the posterior pituitary, in contrast to the brain, is not equipped with an enzyme system to process AVP-(1-9) into its C-terminal metabolites. This may be physiologically meaningful since the metabolites have strong central activities, but are devoid of peripheral endocrine activities, thus demanding production in the brain only.

VI. CONCLUDING REMARKS

Studies on the *in vitro* proteolytic conversion of the natural substrates AVP and OXT by proteolytic enzymes in cellular fractions likely to meet these substrates *in situ*, e.g. synaptic membranes, have successfully led to the identification of novel neuropeptides derived from AVP and OXT.

These metabolites have remarkable biological properties in that they exclusively exert activity in the brain, but not in the periphery, and that they exceed their respective parent peptide several orders of magnitude in potency with respect to affecting consolidation of passive avoidance behavior. Preliminary experiments indicate that these [<Glu^{4},Cyt^{6}]-(4-9) fragments of AVP and OXT are more selective in their central activities than the nonapeptides themselves. For instance whereas AVP-(1-9) normalized the endotoxin induced hyperthermia in the rat, [<Glu^{4},Cyt^{6}]AVP-(4-9) was inactive (Burbach et al., 1983a). Recently, we found that the [<Glu^{4}]-(4-9) metabolites of AVP and OXT do not have a similar highly potent effect on memory retrieval as they have on memory consolidation (in preparation). Therefore, [<Glu^{4}, Cyt^{6}]AVP-(4-9) and [<Glu^{4},Cyt^{6}]OXT-(4-9), or their des-Gly-NH_2 derivatives, may represent the active principles responsible for the facilitating and attenuating effects of AVP and OXT, respectively, on memory consolidation. Other parts of the nonapeptide structures are likely to carry the effects on memory retrieval. The detection of peptides with immunological and chromatographic properties indistinguishable from these metabolites indicates the endogenous presence of these highly active neuropeptides in the brain.

ACKNOWLEDGMENTS

The authors are grateful to Dr. Jan W. van Nispen and Drs. Henk M. Greven for preparation and gifts of synthetic peptides, to Mrs. Joke Cox-van Put for blood pressure measurements, to Miss Patty Zwollo for performing the rat uterus bioassay, to Mrs. Henny de Vos Burchart-Lodewijks for cannulation of the animals, and to Drs. Jos L.M. Lebouille and Drs. Jeroen A. ten Haaf for their contribution to some experiments.

REFERENCES

Ader, R. and De Wied, D., *Psychon. Sci.* 29, 46 (1972).
Boer, G.J., Swaab, D.F., Uylings, H.B.M., Boer, K., Buijs, R.M., and Velis, D.N., *Progr. Brain Res.* 53, 202 (1980).
Bohus, B., *Horm. Behav.* 8, 52 (1977).
Buijs, R.M., *Cell Tiss. Res.* 192, 423 (1978).
Burbach, J.P.H., and De Wied, D., *in* "Neurohypophyseal Peptide Hormones and Other Biologically Active Peptides" (D.H.

Schlesinger, ed.), p. 69. Elsevier North-Holland Inc., Amsterdam (1981).

Burbach, J.P.H., and Lebouille, J.L.M., *J. Biol. Chem.* 258, 1487 (1983).

Burbach, J.P.H., De Kloet, E.R., and De Wied, D., *Brain Res.* 202, 401 (1980a).

Burbach, J.P.H., Schotman, P., Verhoef, J., De Kloet, E.R., and De Wied, D., *Biochem. Biophys. Res. Comm.* 97, 995 (1980b).

Burbach, J.P.H., Wang, X-C., and Van Ittersum, M., *Biochem. Biophys. Res. Comm.* 108, 1165 (1982).

Burbach, J.P.H., Kovács, G.L., De Wied, D., Van Nispen, J.W., and Greven, H.M., *Science* in press (1983a).

Burbach, J.P.H., Bohus, B., Kovács, G.L., Van Nispen, J.W., Greven, H.M., and De Wied, D., *Eur. J. Pharmacol.* in press (1983b).

Cooper, K.E., Kasting, N.W., Lederis, K., and Veale, W.L., *J. Physiol.* 260, 647 (1979).

Dekanski, J., *Br. J. Pharmacol.* 7, 567 (1952).

De Wied, D., *Life Sci.* 19, 685 (1976).

De Wied, D., *Proc. R. Soc. Lond. B* 210, 183 (1980).

De Wied, D., and Bohus, B., *Progr. Brain Res.* 48, 327 (1978).

De Wied, D., Greven, H.M., Lande, S., and Witter, A., *Br. J. Pharmacol.* 45, 118 (1972).

Ferrier, B.M., Kennett, D.J., and Devlin, M.C., *Life Sci.* 27, 2311 (1980).

Jolles, J., *Progr. Brain Res.* in press (1983).

Pedersen, C.A., and Prange, A.J., *Proc. Natl. Acad. Sci. USA* 76, 6661 (1979).

Södersten, P., Henning, M., Melin, P., and Ludin, S., *Nature* 301, 608 (1983).

Van Nispen, J.W., Hannink, J.A.J., and Greven, H.M., *in* "Peptides 1983", IXth American Peptide Symposium, in press (1983).

Van Ree, J.M., and De Wied, D., *in* "Endogenous Peptides and Learning and Memory Processes" (J.L. Martinez Jr., et al., eds.), p. 397. Academic Press, New York (1981).

Wang, X-C., Burbach, J.P.H., Verhoef, J., and De Wied, D., *Brain Res.* in press (1983).

Wang, X-C., Burbach, J.P.H., Ten Haaf, J.A., and De Wied., D., submitted (1983).

Weindl, A., and Sofroniew, M.V., *in* "Brain and Pituitary Peptides" (W. Wuttke, A. Weindl and K.H. Voight, eds.), p. 97. Karger, Basel (1980).

PART VI

Structure and Biosynthesis of Releasing Factors

CHARACTERIZATION OF RAT HYPOTHALAMIC GRCWTH HORMONE AND CORTICOTROPIN RELEASING FACTORS

Joachim Spiess
Jean Rivier
Wylie Vale

Peptide Biology Laboratory
The Salk Institute
La Jolla, California

I. INTRODUCTION

More than thirty years ago, it was proposed (Harris, 1948; Green and Harris, 1949) that the hypothalamic hypophyseal portal blood vessel system was of crucial importance in the hypothalamic control of the anterior pituitary. It was thought that factors produced in the hypothalamus would be released into portal vessels to be transported by the bloodstream to the anterior pituitary and to modify hormonal secretion. Based on this hypothesis, three hypothalamic hypophysiotropic peptides containing 3-14 residues were characterized from 1968-1972. These peptides were the tripeptide thyrotropin-releasing hormone (TRH) (Burgus et al., 1970; Nair et al., 1970), the decapeptide gonadotropin-releasing hormone (GnRH) (Matsuo et al., 1971; Burgus et al., 1972) and the tetradecapeptide growth hormone (GH) release inhibiting factor (GRIF), somatostatin-14 (Brazeau et al., 1973; Burgus et al., 1973). Later, another (larger) form of GRIF, somatostatin-28, was discovered (Pradayrol et al., 1980; Schally et al., 1980; Esch et al., 1980; Spiess et al., 1981a).

This research was supported in part by NIH grants AM-26741, AM20971 and HD13527, and by grants from The Rockefeller Foundation and The March of Dimes-Birth Defects Foundation. Sequence analysis facilities were supported in part by the Stiftung Volkswagenwerk and Fritz Thyssen Stiftung. Research was conducted in part by The Clayton Foundation for Research, California Division. J.S. and W.V. are Clayton Foundation Investigators.

ISBN 0-12-417320-9

1 5 10 15
H-Tyr-Ala-Asp-Ala-Ile-Phe-Thr-Asn-Ser-Tyr-Arg-Lys-Val-Leu-Gly-

16 20 25 30
-Gln-Leu-Ser-Ala-Arg-Lys-Leu-Leu-Gln-Asp-Ile-Met-Ser-Arg-Gln-

31 35 40
-Gln-Gly-Glu-Ser-Asn-Gln-Glu-Arg-Gly-Ala-Arg-Ala-Arg-Leu-NH_2

FIGURE 1. Primary structure of hpGRF-44 and the shorter forms hpGRF-40 and hpGRF-37. Both human pancreatic tumors investigated (Rivier, J. et al., 1982; Guillemin et al., 1982) contained hpGRF-40, whereas hpGRF-44 and hpGRF-37 were only found in one of the tumors (Guillemin et al., 1982).

In the past two years, several large highly potent hypophysiotropic peptides have been biologically and chemically analyzed: ovine hypothalamic corticotropin-releasing factor (ohCRF) (Vale et al., 1981; Spiess et al., 1981b), and human pancreatic tumor growth hormone releasing factors (hpGRF's) (Guillemin et al., 1982; Spiess et al., 1982a; Rivier, J. et al, 1982) (Fig. 1).

The 41-residue polypeptide ohCRF was the first hypothalamic CRF of high intrinsic corticotropin (ACTH) releasing activity, the primary structure of which was completely elucidated. The significance of ohCRF or a structurally closely related peptide as physiologic regulator of the ACTH secretion has been demonstrated by the stimulation of the hypophyseal ACTH secretion with ohCRF in vitro and in vivo (Vale et al., 1981; Rivier, C. et al., 1982a) by the finding of a CRF concentration in portal blood in the range of the half-maximally effective ohCRF concentration (Gibbs and Vale, 1982), by the immunohistochemical distribution of CRF in the brain (Bloom et al., 1982; Swanson et al., 1983) and by the lowering of plasma ACTH levels after administration of ohCRF antibody prior to ohCRF application or stress exposure (Rivier, C. et al., 1982b).

It was observed that human, rat and dog hypothalamic extracts exhibited significantly higher immunoactivity in radioimmunoassay (RIA) C-24 directed against ohCRF(4-20) than in RIA's directed versus the middle or C-terminal sequence of ohCRF (Vale et al., 1983a). This observation, which was suggestive for partial sequence differences between these peptides, could reflect species variation or the presence of multiple forms of CRF in the same species or both. The existence of multiple forms of CRF could especially not be excluded, because the ohCRF characterized by us was purified from a side fraction which contained only a small portion of the total ACTH releasing activity (Vale et al., 1981), and thus could not conclusively be considered as the main CRF of ovine hypothalamus. We therefore considered it important to purify CRF from fresh tissues and to determine its primary structure. Rat was chosen as the source of hypothalamic tissue to facilitate studies on CRF in the principal laboratory animal. Rat hypothalamic extracts served at the same time as the source for hypothalamic GRF, which behaved like rat hypothalamic CRF (rhCRF) in the initial steps of purification.

Most details of the characterization of rat hypothalamic GRF (rhGRF) and rhCRF have been described elsewhere (Spiess et al., 1983a,b; Rivier, J. et al., 1983). We present here additional aspects.

II. EXPERIMENTAL PROCEDURES

A. *Purification*

rhCRF and rhGRF were extracted (Vale et al., 1983b) with aqueous acetic acid containing β-mercaptoethanol and protease inhibitors from the lyophilisate of 20,000-60,000 rat hypothalami, provided by Dr. A. Parlow under the aegis of NIADDK. The subsequent purification with gel filtration and reverse-phase high-pressure liquid chromatography (HPLC) was monitored with ohCRF RIA C-24 and the rat anterior pituitary cell culture assay (Spiess et al., 1983a; Rivier, J. et al., 1983).

B. *Edman Degradation*

Approximately 0.4 - 1.4 nmol of peptide, in some experiments pretreated with 3-sulfo-phenylisothiocyanate, was applied to a modified (Wittmann-Liebold, 1980) Beckman 890C spinning cup sequencer and degraded as described (Spiess et al., 1982b).

C. *Peptide Mapping*

Peptides (0.1 - 1.5 nmol) were digested either with clostripain (Boehringer, Mannheim) or staphylococcal protease (Miles) and then subjected to reverse-phase HPLC, which was performed on large pore Vydac C_{18} columns eluted with mixtures of aqueous trifluoroacetic acid and acetonitrile. The column eluate was fractionated and analyzed by amino acid analysis and Edman degradation. The conditions for the mapping of each native peptide were worked out in a series of kinetic experiments with synthetic replicates of the native peptides under investigation. Details are described elsewhere (Spiess et al., 1983a,b).

D. *Peptide Synthesis*

Peptides were synthesized by the solid-phase approach (Merrifield, 1963) modified as described (Rivier, 1974; Märki et al., 1981). The peptides cleaved from the solid phase were purified mainly with reverse-phase HPLC and characterized like the native peptides with amino acid analysis, peptide mapping, Edman degradation and bioassays.

III. RESULTS AND DISCUSSION

Rat hypothalamic CRF and GRF were purified from the same starting material with similar procedures. The final purification accomplished in separate experiments with reverse-phase HPLC yielded approximately 90 fmol of rhCRF and 40 fmol of rhGRF per hypothalamus. Based on the analysis of all column eluates with radioimmunoassay or bioassay, the purified peptides represented the main respective forms of high intrinsic ACTH and GH releasing activities of the rat hypothalamic extract.

The strategies of sequence analysis of rhCRF and rhGRF were similar. Most sequence information was obtained by Edman degradation of 0.9 - 1.4 nmol of peptide. This sequence information was supplemented and confirmed by peptide mapping on reverse-phase HPLC and bioassays of the native peptides and their synthetic replicates.

One important difference between the sequence analyses of rhGRF and rhCRF was that Edman degradation of rhGRF was accomplished with a peptide of a purity of 90-95%, whereas the rhCRF preparation applied to Edman degradation was found to be of a purity of approximately 50%. Sequence analysis of partially purified rhCRF was possible on the basis of the orthophthalaldehyde (OPA) strategy which allows Edman degradation of partially purified peptides containing proline residues under quasi-homogeneous conditions. Contaminating peptides are blocked with OPA in Edman cycles in which the peptide of interest carries an N-terminal proline residue not affected by OPA. Details are described elsewhere (Spiess et al., 1983b,c).

This strategy may not only be useful in the characterization of CRF (of a purity of 5-10%) from other sources, but generally applicable to partial sequence analysis of proline containing peptides which are difficult to purify. Sequence information obtained under these conditions could be valuable for the production of antibodies or the synthesis of oligonucleotide probes.

rhCRF was determined to be a 41-residue polypeptide C-terminally blocked by an amide group. The sequence homology of rhCRF with ohCRF was calculated to be 83%. The seven amino acid differences (Fig. 2) which represent chemically conservative replacements could be explained, with the exception of the C-terminal difference, as the result of single base mutations. Both peptides did not differ significantly in their intrinsic activities and potencies _in vitro_ and _in vivo_ in the rat (Rivier, J. et al., 1983; Rivier, C., in preparation).

Recently, the primary structure of human CRF has been predicted from the nucleotide sequence of a human genomic DNA segment (Shibahara et al., 1983). This predicted sequence is identical with the sequence of rhCRF.

To date, five CRF-like polypeptides have been characterized. These peptides are mammalian ohCRF, rhCRF, amphibian skin sauvagine from the frog _Phyllomedusa sauvagei_ (Montecucchi et al., 1979), and bony fish urophyseal urotensin I from _Catastomus commersoni_ and _Cyprinus carpio_ (Lederis et al., 1982). A comparison of the primary structures of rhCRF and the known non-mammalian CRF-like peptides reveals sequence homologies of 54% with urotensin I and 45% with sauvagine

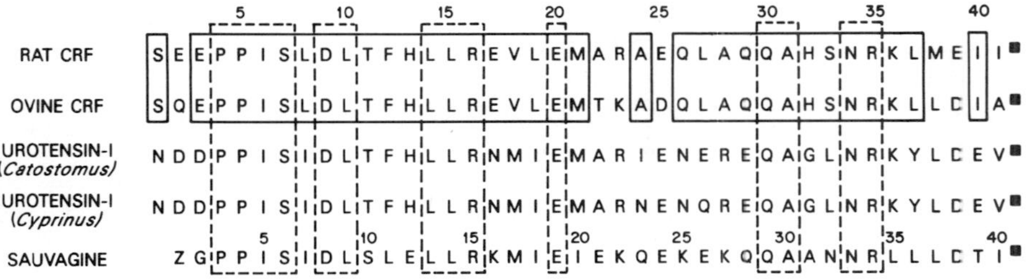

FIGURE 2. Sequence homologies of hypothalamic CRF's and CRF-like polypeptides. Alignments are marked by rectangles. ____, comparison of rat and ovine CRF. - - -, residues homologous to all members of this peptide family. A, alanine; D, aspartic acid; E, glutamic acid; F, phenylalanine; G, glycine; H, histidine; I, isoleucine; K, lysine; L, leucine; M, methionine; N, asparagine; Q, glutamine; R, arginine; S, serine; T, threonine; V, valine; Y, tyrosine; Z, pyrrolidonecarboxylic acid; ■, amidated COOH-terminus.

(Fig. 2). Whether the sequence differences between mammalian and non-mammalian CRF's reflect species or tissue variations remains unclear as long as hypothalamic CRF-like peptides from the non-mammalian species mentioned above have not been characterized.

It is interesting that all CRF-like peptides are not only structurally closely related, but also exhibit statistically equivalent hypophysiotropic activities (Rivier, C. et al., 1983) and exert similar extra-hypophyseal actions (Brown, M. et al., 1982; Lenz et al., 1983). These peptides thus constitute a peptide family (the CRF urotensin I sauvagine family) on the basis of their close structural and biological relationship.

rhGRF was found to be a 43-residue polypeptide with free C-terminus (Fig. 3). The determination of the primary structure of this peptide represents the first successful chemical characterization of a GRF of established hypothalamic origin. The sequences of rhGRF and hpGRF-44, the largest of the three hpGRF's discovered in extracts of human pancreatic tumors, align with 29 residues (Fig. 3). This alignment corresponds to a sequence homology of 67%. In the rat anterior pituitary cell culture assay, rhGRF appears to be slightly more potent than hpGRF-44 in stimulating GH secretion. Under the same conditions, rhGRF and hpGRF are significantly more potent than hpGRF-40 (Spiess et al., 1983a).

It has become increasingly clear that the differences between hpGRF and rhGRF represent species and not tissue variation. Recently, two forms of GH releasing activity have been extracted from 5000 human hypothalami (provided by Dr. Seymour Reichlin) and purified mainly with reverse-phase HPLC. Final purification was accomplished only after oxidation of both peptides to the respective methionine sulfoxide forms. These chemically modified peptides coeluted in reverse-phase HPLC with the oxidized forms of synthetic hpGRF-44 and hpGRF-40, respectively. On the basis of peptide mapping on reverse-phase HPLC after digestion of 100-150 pmol of the methionine sulfoxide forms of the two native peptides with clostripain,

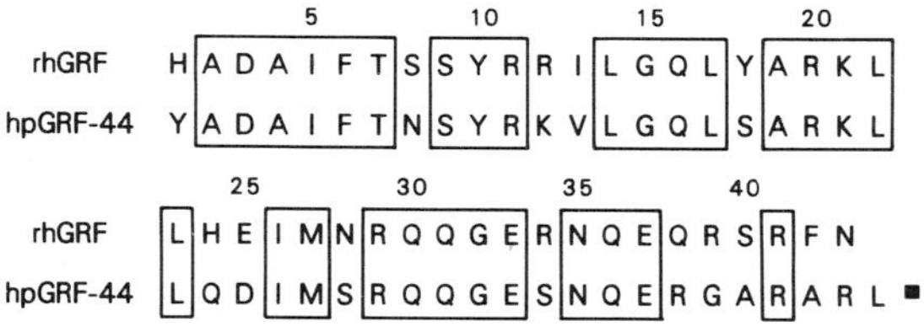

FIGURE 3. Sequence comparison of rhGRF and hpGRF-44. Alignments are indicated by rectangles.

it was indicated by retention times and amino acid analysis that both native peptides contained the N-terminal sequence of hpGRF (1-29) (Spiess et al., 1983a) observed to be responsible for the full intrinsic activity of hpGRF (Rivier, J. et al., 1982). Although the data of reverse-phase HPLC of the two native human hypothalamic GRF's (hhGRF's) and their digests are strongly suggestive for structural identity of hhGRF and hpGRF, final proof has to be provided by sequence analysis of the purified native peptides.

If one assumes identity of hhGRF and hpGRF, as well as of hhCRF and rhCRF, it is suggested that CRF may be more conserved in the peptide evolution than GRF.

On the basis of structural comparison, rhGRF belongs to the glucagon secretin GRF family, which consists of porcine PHI-27 (Tatemoto and Mutt, 1981), vasoactive intestinal peptide (VIP) (Mutt and Said, 1970), and gastric inhibitory peptide (GIP) (Brown, J.C. and Dryburgh, 1971), hpGRF and rhGRF. Although the sequence homologies of rhGRF with PHI-27, VIP, glucagon, secretin and GIP are 52, 43, 31, 30 and 14%, respectively, only rhGRF and hpGRF exhibit significant GH releasing activity. The structural relationship between the peptides of this family is thus only in part matched with similarities in biological actions. In this respect, the glucagon secretin GRF family contrasts with the CRF urotensin I sauvagine family.

An interesting aspect of primary structure comparisons is the multiplicity of hhGRF and hpGRF. In view of the observation mentioned above that hpGRF-44 is biologically more potent than hpGRF-40, it seems possible that this multiplicity be of regulative significance. However, it cannot be excluded at this time that the observed multiplicity results from artefactual cleavages during extraction and purification.

ACKNOWLEDGMENTS

We thank Dr. A. Parlow and The National Hormone and Pituitary Program, and Dr. S. Reichlin for providing hypothalamic tissue; J. Heil, R. Lee, J. Vaughan, R. Galyean, G. Yamamoto, R. Kaiser and R. McClintock for technical assistance; and D. Weigand and S. McCall for manuscript preparation.

REFERENCES

Bloom, F. E., Battenberg, E., Rivier, J., and Vale. W., *Reg. Peptides 4*, 43 (1982).

Brazeau, P., Vale, W., Burgus, R., Butcher, M., Rivier, J., and Guillemin, R., *Science 179*, 77 (1973).

Bromer, W. W., Sinn, L. G., and Behrens, O. K., *J. Amer. Chem. Soc. 79*, 2807 (1957).

Brown, J. C. and Dryburgh, J. R., *Canad. J. Biochem. 49*, 867 (1971).

Brown, M., Fisher, L. A., Spiess, J., Rivier, J., Rivier, C., and Vale, W., *Regulatory Peptides 4*, 107 (1982).

Burgus, R., Dunn, T. F., Desiderio, D., Ward, D. N., Vale, W., and Guillemin, R., *Nature 226*, 321 (1970).

Burgus, R., Butcher, M., Amoss, M., Ling, N., Monahan, M., Rivier, J., Fellows, R., Blackwell, R., Vale, W., and Guillemin, R., *Proc. Natl. Acad. Sci. U.S.A. 69*, 278, (1972).

Burgus, R., Ling, N., Butcher, M., and Guillemin, R., *Proc. Natl. Acad. Sci. U.S.A. 70*, 684 (1973).

Esch, F., Böhlen, P., Ling, N., Benoit, R., and Guillemin, R., *Proc. Natl. Acad. Sci. U.S.A. 77*, 6827 (1980).

Gibbs, D. M., and Vale, W., *Endocrinology 111*,1418 (1982).

Green, J. D., and Harris, G. W., *J. Physiol. (London) 108*, 359 (1949).

Guillemin, R., Brazeau, P., Böhlen, P., Esch, F., Ling, N., and Wehrenberg, W. B., *Science 218*, 585 (1982).

Harris, G. W., *Physiol. Rev. 28*, 139 (1948).

Lederis, K., Letter, A., McMaster, D., Moore, G., and Schlesinger, D., *Science 218*, 162 (1982).

Lenz, H. J., Fisher, L. A., Rivier, J., Vale, W., and Brown, M. R., *Gastroenterology 84*, 1228 (1983) (abstract).

Märki, W., Spiess, J., Taché, Y., Brown, M., and Rivier, J. E., *J. Amer. Chem. Soc. 103*, 3178 (1981).

Matsuo, H., Baba, Y., Nair, R. M. G., Arimura, A., and Schally, A. V., *Biochem. Biophys. Res. Commun. 43*, 1374 (1971).

Merrifield, R. B., *J. Amer. Chem. Soc. 85*, 2149 (1963).

Montecucchi, P. C., Henschen, A., and Erspamer, V., *Hoppe-Seyler's Z. Physiol. Chem. 360*, 1178 (1979) (abstract).

Mutt, V., Jorpes, J. E., and Magnusson, S. D., *Eur. J. Biochem. 15*, 513 (1970).

Mutt, V., and Said, S. I., *Eur. J. Biochem. 42*, 581 (1974).

Nair, R. M. G., Barrett, J. F., Bowers, C. Y., and Schally, A. V., *Biochemistry 9*, 1103 (1970).

Pradayrol, L., Jörnvall, H., Mutt, V., and Ribet, A., *FEBS LETT. 109*, 55 (1980).

Rivier, C., Brownstein, M., Spiess, J., Rivier, J., and Vale, W., *Endocrinology 110*, 272 (1982a).
Rivier, C., Rivier, J., and Vale, W., *Science 218*, 377 (1982b).
Rivier, C., Rivier, J., Lederis, K., and Vale, W., *Regulatory Peptides 5*, 139 (1983).
Rivier, J., *J. Amer. Chem. Soc. 96*, 2986 (1974).
Rivier, J., Spiess, J., Thorner, M., and Vale, W., *Nature 300*, 276 (1982).
Rivier, J., Spiess, J., and Vale, W., *Proc. Natl. Acad. Sci. U.S.A., in press* (1983).
Schally, A. V., Huang, W.-Y., Chang, R. C., Arimura, A., Redding, T. W., Millar, R. P., Hunkapiller, M. W., and Hood, L. E., *Proc. Natl. Acad. Sci. U.S.A. 77*, 4489 (1980).
Shibahara, S., Morimoto, Y., Furutani, Y., Notake, M., Takahashi, H., Shimizu, S., Horikawa, S., and Numa, S., *EMBO Journal 2*, 775 (1983).
Spiess, J., Villarreal, J., and Vale, W., *Biochemistry 20*, 1981 (1981a).
Spiess, J., Rivier, J., Rivier, C., and Vale, W., *Proc. Natl. Acad. Sci. U.S.A. 78*, 6517 (1981b).
Spiess, J., Rivier, J., Thorner, M., and Vale, W., *Biochemistry 24*, 6037 (1982a).
Spiess, J., Rivier, J., Rivier, C., and Vale, W., *in* "Methods in Protein Sequence Analysis" (M. Elzinga, ed.), p. 131, Humana Press, Clifton, New Jersey (1982b).
Spiess, J., Rivier, J., and Vale, W., *Nature 303*, 532 (1983a).
Spiess, J., Rivier, J., and Vale, W., *Biochemistry, in press* (1983b).
Spiess, J., *in* "Proceedings of the Eighth American Peptide Symposium" (V. J. Hruby and D. H. Rich, eds.), in press, Pierce Chemical Company, Rockford, Illinois (1983c).
Swanson, L., Sawchenko, P., Rivier, J., and Vale, W., *Neuroendocrinology 36*, 165 (1983).
Tatemoto, K., and Mutt, V., *Proc. Natl. Acad. Sci. U.S.A. 78*, 6603 (1981).
Vale, W., Spiess, J., Rivier, C., and Rivier, J., *Science 213*, 1394 (1981).
Vale, W., Rivier, C., Spiess, J., and Rivier, J., *in* "Brain Peptides" (D. Krieger, M. Brownstein, and J. Martin, eds.), in press, John Wiley & Sons, New York, (1983a).
Vale, W., Vaughan, J., Yamamoto, G., Spiess, J., and Rivier, J., *Endocrinology 112*, 1553 (1983b).
Wittmann-Liebold, B., *in* "Polypeptide Hormones" (R. F. Beers, Jr. and E. G. Bassett, eds.), p. 87, Raven Press, New York (1980).

PURIFICATION STUDIES ON PUTATIVE LHRH PRECURSORS

Jean-Pierre Gautron
Eliane Pattou
Claude Kordon

Unité de Neuroendocrinologie
INSERM
Paris, France

I. INTRODUCTION

The occurence of LHRH precursor forms of larger molecular weight has been postulated for a long time. Subcellular fractionation studies had shown that most (around 90%) LHRH immunoreactivity was recovered from synaptosomal fractions containing almost exclusively nerve endings. It was thus indirectly concluded that the initial biosynthesis products present in the perikaryal cytoplasm were probably not, or only poorly recognized by antibodies directed against the native or the synthetic peptide (Ramirez et al., 1975). In line with this finding was the observation that cytoimmunochemical labelling of LHRH cell bodies permitted detection of a very limited number of neurons only (Barry, 1978), a situation very different to that observed for oxytocin, vascpressin or neurophysin cell bodies (Vandesande and Dierickx, 1975) for instance.

An additional difficulty in attempting to identify precursor forms of the hormone is due to the scarcity of LHRH producing cells. The peptide is not as ubiquitous in the brain and the organism as most other peptide transmitters or neurohormones ; the main LHRH projecting system in the brain is entirely contained within the hypothalamus. Very scarce amounts of immunoreactive-like material have also been described in the placenta (Siler-Khodr and Khodr, 1978), in a sympathetic ganglion (Jan et al., 1979 ; Eiden and Eskay, 1980) and in the

BIOCHEMICAL AND CLINICAL ASPECTS OF
NEUROPEPTIDES: SYNTHESIS, PROCESSING,
AND GENE STRUCTURE

ISBN 0-12-417320-9

gonads (Dutlow and Millar, 1981), where binding sites for the peptides have also been found (Clayton et al., 1980 ; Heber et al., 1978). It is not clear yet whether this immunoreactivity corresponds to a peptide structure identical to that of hypothalamic LHRH. No other organ has ever been found to contain the neuropeptide, whether under normal or under pathological conditions.

As a result of this, characterization of LHRH precursor forms based on translation and identification of mRNA coding for the peptide is extremely difficult, since brain mRNA coding for the peptide represents a very small minority of total mRNA (possibly less than 1 in 10000) (Richter, 1983). Consequently, most investigators who attempted to translate LHRH precursors in vitro failed to obtain it or ended up with very small amounts of material only (Curtis and Fink, 1983).

Two additional difficulties impair elucidation of LHRH precursors. Firstly, the decapeptide appears to have a very labile conformation ; it is very susceptible to structural changes, to pH or to solvent molarity, as shown by circular dichroism studies (Marche et al., 1976). This may account for the very poor activity exhibited by antibodies directed at the inner structure of the peptide. Even so-called conformational antibodies loose much of their activity in the presence of peptides with N- or C-terminal modifications ; this will be illustrated in the next part of this chapter. With most (and, in our hands, with all) available antibodies, incubation with precursor forms thus leads to a severe underevaluation of the material. Second, this conformational lability and the presence of two aromatic aminoacids in closely related position of the molecule causes an abnormal behaviour of the neuropeptide and of its N-terminal fragments upon molecular sieve filtration, as will also be illustrated in the next section ; abnormal coelution of larger molecular forms and catabolites of LHRH are thus liable to generate artefacts.

In the present chapter, we will illustrate purification studies carried out on a putative intermediate precursor form of LHRH, based on the sequential use of several fractionation techniques : molecular sieve filtration on various gels, separation on ion exchange resins, and affinity chromatography with insolubilized antibodies directed against various parts of the neuropeptide.

II. PURIFICATION OF AN INTERMEDIATE LHRH PRECURSOR

The first step of the work involved synthesis of extended LHRH sequences in order to test the capacity of several antibodies to recognize the peptide sequence when it is included in a larger protein. An Arg-Lys or an acetyl-Arg-Lys sequence was added to the N-terminal end of the peptide ; a Gly-Ala-NH_2 was added at the amidated C-terminal of the peptide. C-terminal extension completely blocked the capacity of antibodies directed against the C-terminal to recognize the fragment. In addition, antibodies described as directed at a more inner part of the peptide structure were also unable to show competition between LHRH and the C-extended sequence (Fig. 1).

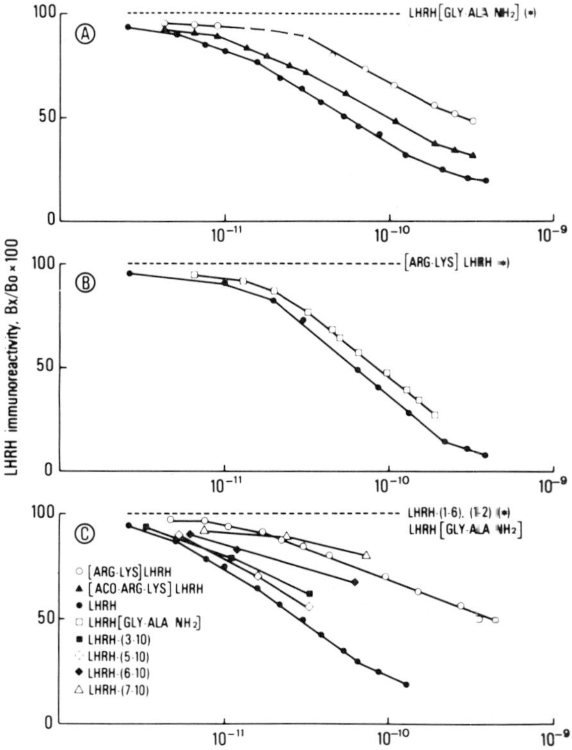

FIG. 1. Effects of increasing molarities of synthetic peptides on the displacement of iodinated LHRH from C (A), N (B) and W (C) antibodies. *For all 3 antibodies, none of the following substances showed cross reactivity with synthetic LHRH : TRH, SRIF, VIP, substance P, bombesin, neurotensin, bradykinin, glucagon, insulin A chain, insulin B chain. Abscissa : Log molecular weight (from Gautron et al., 1981).

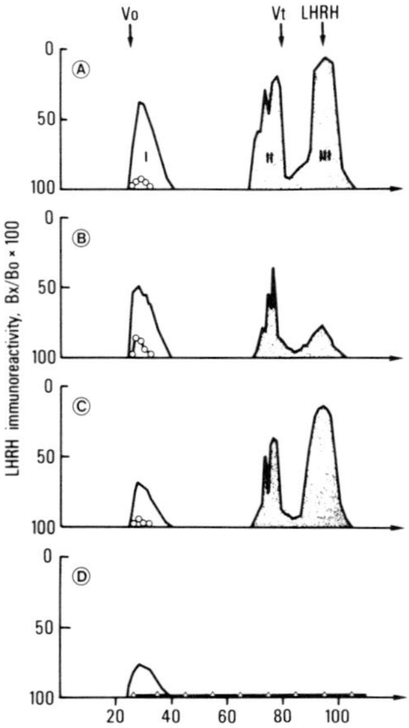

FIG. 2. Fractionation by molecular sieve filtration of LHRH and LHRH-like immunoreactivities (N antibody) from hypothalamic subcellular fractions. Extracts (A, homogenate ; B, supernatant S_2 ; C, crude synaptosomal pellet P_2) were fractionated on a 100 cm x 1.6 cm Bio-Gel P_4 column. On (D), - and Δ represent, respectively, apparent immunoreactivities after fractionation of bacitracin lyophilized in 2 N AcOH in the presence of absence of triton X-100. On (A, B and C) - and 0 for peak I represent, respectively, immunoreactivities from subcellular fractions extracted with 2 N AcOH in the presence or absence of triton X-100.

This is consistent with observations that so-called inner directed antibodies loose much of their affinity towards sequences bearing only slight C-terminal modifications, for instance acid(deaminated) LHRH (Copeland et al., 1979 ; Millar et al., 1978). All antibodies we tested, whether directed against the C, the N or the inner part of the molecule, showed some degree of cross-reaction towards LHRH fragments containing respectively the N-terminal of the C-terminal sequence.

As shown on Fig. 2, immunoreactive material was recovered from the void volume and in two distinct elution fractions of a Bio-Gel P_4. After subcellular fractionation, the heaviest material, which remains in the void volume, and the first elution peak were found to be predominant in cytoplasmic

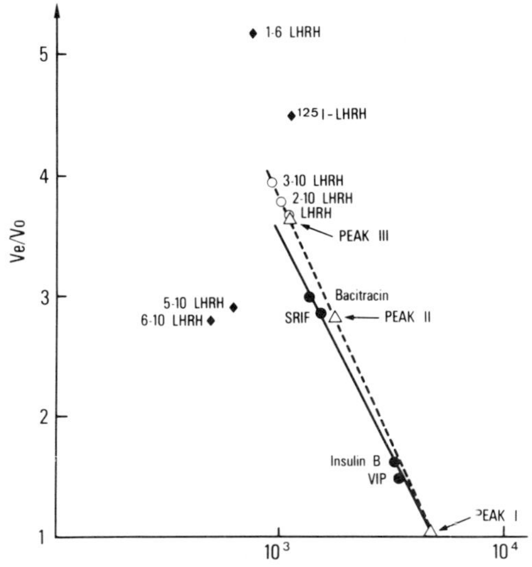

FIG. 3. *Chromatography of substances of known molecular weight on Bio-Gel P_4. The ratio of elution volume (V_e) to void volume (V_o) of the various substances are plotted against the log of their molecular weights. Substances were chromatographed on a 100 cm x 1.6 cm Bio-Gel P_4 column in 0.2 N acetic acid containing 0.1% triton X-100, 2×10^{-4}M bacitracin and 0.02% sodium azide. The flow rate was 5 ml/h. The fraction size was 2.5 ml. From Gautron et al., 1981.*

fractions, whereas native LHRH was mostly recovered from a synaptosomal fraction. At that stage, we concentrated our efforts on purification of the intermediate peak.

Fig. 3 shows the elution conditions of that fraction, as compared to several polypeptide markers. Given its molecular weight, LHRH and several related sequences, in particular fragments 1-8 or 1-9, appear retarded on the column as compared to insulin, somatostatin or similar peptide markers. This retarded behavior can be attributed to the effect of the aromatic aminoacids in position 3 and 5 inasmuch as peptide fragments which do not contain these aminoacids, such as the C-terminal fragments,are not affected. When read against a straight line corrected for this retardation effect, the small peak II precursor of LHRH was eluted in a position corresponding to about 1800 daltons.

In addition, Fig. 3 shows that smaller C-terminal catabolites of LHRH can coelute with the material present as peak II. Since these smaller peptides can be recognized by C-terminal antibodies, and, to a smaller extent, by so-called conformational antibodies as well, peak II eluate was

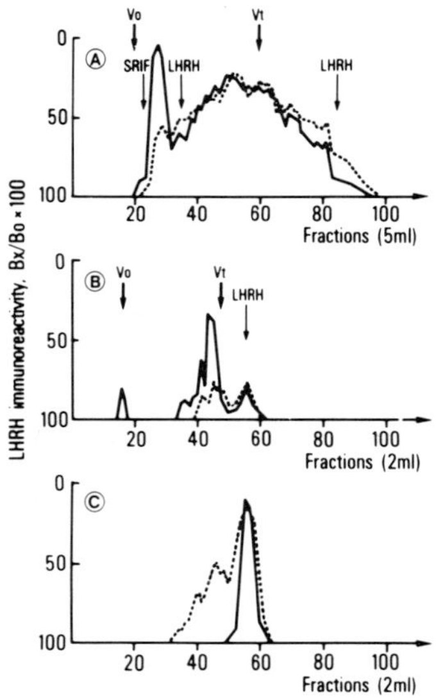

FIG. 4. Chromatographic behavior of hypothalamic LHRH and LHRH-like immunoreactivities (—— N antibody ; --- C antibody) on different gels. (A) tissue was fractionated on a 150 cm x 1.6 cm Sephadex G-10 column in 0.2 N AcOH containing 0.02% sodium azide. Synthetic SRIF was eluted as a single peak ; synthetic LHRH was fractionated between the two arrows. (B) tubes 20-32 and (C) tubes 33-80 from the G-10 column were pooled and fractionated on a 100 cm x 1.1 cm Bio-Gel P_4 column in 50 mM NHAc buffer, pH 4.2, containing 0.1% triton X-100, 2×10^{-4}M bacitracin and 0.02% sodium azide.

rechromatographed and assayed against antibodies of different specificities (Fig. 4). After these experiments, it was found that a part of the immunoreactivity of peak II could be accounted for by contamination of cleavage products. This contamination accounts for the immunoreactivity found with C-terminal antibodies. In contrast, immunoreactive material found to react to the N-terminal antibody can only be accounted for by a larger fragment. We were thus able to conclude that the 1800 dalton material should be attributed to a prolongation of LHRH on the carboxyterminal (10 Gly).

The material eluted from peak II was then submitted to a second purification step using ion exchange resins. To our surprise, the putative precursor was not retained on a cation exchange resin, under conditions which permit fixation of LHRH itself. In contrast, it was retained on an anionic exchange column and could be eluted by low molarities of NH_4 acetate.

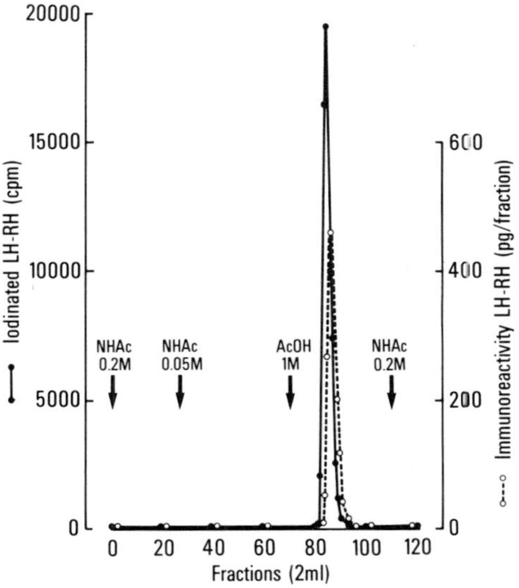

FIG. 5. Affinity chromatography of putative precursors of LHRH from rat hypothalamus using an N-terminal antibody bound to Sepharose 4B. LHRH immunoreactivity was cycled through the column in 0.2 M ammonium acetate (pH 7.0) for 6h, non-immunoreactive material was washed off with 0.2 M and 0.05 M ammonium acetate and the putative precursor was eluted with 1 N acetic acid.

The immunoreactive precursor was purified about 800-fold in the eluate of the anionic exchange column.

After lyophilization, the eluate of the anionic exchange column was repurified on an affinity chromatography column, on which antibodies against the N-terminal fragment of LHRH have been insolubilized. The material could be eluted from the column by weak acidic buffers. It could be shown to co-elute with synthetic LHRH (Fig. 5). The extent of purification in the eluate of the affinity column is not easy to calculate precisely, because the total protein content of that fraction is hardly measurable ; it can however be estimated at or above 10,000-fold.

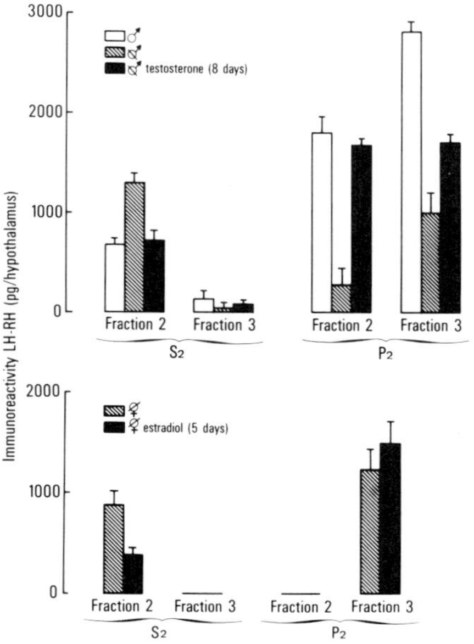

FIG. 6. *Subcellular distributions of putative precursor (fraction 2) and LHRH (fraction 3) in the rat hypothalamus after castration and after restoring steroid feedback conditions by administration of testosterone or estradiol.*

III. PHYSIOLOGICAL STUDIES

So far, our postulate that the extended LHRH sequence recovered from the hypothalamus and recognized by N-terminal antibodies is indeed an intermediate precursor of the peptide relies only on indirect data. Final demonstration of the hypothesis will only be possible when the peptide will be fully sequenced. In the meantime, a few physiological observations have lend support to the presumption of a relationship between the extended peptide and LHRH.

Both LHRH and the 1800 dalton immunoreactive form exhibit characteristic subcellular distribution patterns (Fig. 6). Only trace amount of the native peptides are recovered from a 14500 g supernatant in sucrose, which corresponds to axoplasmic or cytoplasmic material (S_2) ; the peptide is mostly concentrated within nerve endings, spinned down as synaptosomes into a mitochondrial fraction (P_2). In contrast, the extended immunoreactive form can be found both in the supernatant and the mitochondrial pellet. Three weeks after castration, a condition known to increase the turnover rate of LHRH (Kordon et al., 1976), the relative proportions of both molecular forms are markedly changed ; the larger molecular form decreases significantly in both fractions prepared from castrated males, and becomes undetectable in castrated females. After restoring steroid feedback conditions by administration of testosterone or estradiol, concentrations of the putative precursor come back to their precastration values. These data are consistent with the hypothesis that the larger immunoreactive form participates in the metabolism of LHRH, and may thus be an intermediate precursor of the peptide.

IV. CONCLUSION

We thus have obtained a highly purified peptide sequence which, after immunological as well as physiological criteria, behave like a tentative intermediate precursor of LHRH. The eluate of affinity chromatography is presently submitted to a final HPLC purification ; we hope to be able to sequence it immediately after. At present, we can only speculate that the extended sequence probably contains a double Gly in the C-terminal region. Characterization of an intermediate precursor with a C-terminal elongated sequence could be very helpful, because it would permit to raise antibodies against a larger sequence of the precursor molecule. This procedure may overcome the poor cross-reactivity exhibited by conventional anti-LHRH antisera against antigenic determinants included within a large protein. As we have seen, this low level of cross reaction is probably due to the conformational instability of the native decapeptide.

REFERENCES

1. Barry, J., *Mol. Cell. Endocrinol.* *3*, 339 (1975).
2. Clayton, R. N., Katikinemi, M., Chan, V., Dufau, M., and Catt, K. J., *Proc. Natl. Acad. Sci., USA*, *77*, 4459 (1980).
3. Copeland, K. C., Aubert, M. L., Rivier, J., and Sizonenko, P. C., *Endocrinology* *104*, 1504 (1979).
4. Curtis, A., and Fink, G., *Endocrinology* *112*, 390 (1983).
5. Dutlow, C. M., and Millar, R. P., *Biochem. Biophys. Res. Commun.* *101*, 486 (1981).
6. Eiden, L. E., and Eskay, R. L., *Neuropeptides* *1*, 29 (1980).
7. Heber, D., Marshall, J. C., and Odell, W. D., *Amer. J. Physiol.* *235*, 227 (1978).

8. Jan, Y. N., Jan, L. Y., and Kuffler, S. W., *Proc. Natl. Acad. Sci., USA, 76,* 1501 (1979).
9. Kordon, C., Epelbaum, J., Enjalbert, A., and McKelvy, J., *in* " Subcellular Mechanisms in Reproductive Neuroendocrinology " (F. Naftolin, K. J. Ryan and J. Davies, Eds.), p. 167. Elsevier Scientific Publishing Company, Amsterdam, (1976).
10. Marche, P., Montenay-Garestier, T., Helene, C., and Fromageot, P., *Biochemistry 15,* 5730 (1976).
11. Millar, R. P., Denniss, P., Tobler, C., King, J. C., Schally, A. V., and Arimura, A., *in* " Biologie Cellulaire des Processus Neurosécrétoires Hypothalamiques " (J. D. Vincent, and C. Kordon, Eds.), p. 487. C.N.R.S., Paris, (1978).
12. Ramirez, V. D., Gautron, J. P., Epelbaum, J., Pattou, E., Zamora, A., and Kordon, C., *Mol. Cell. Endocrinol. 3,* 339 (1975).
13. Richter, D., *in* this volume, in press.
14. Siler-Khodr, T. M., and Khodr, G., *Amer. J. Obstet. Gynec. 130,* 216 (1978).
15. Vandesande, F., and Dierickx, F., *Cell. Tissue Res. 164,* 153 (1975).

TENTATIVE IDENTIFICATION OF LUTEINIZING HORMONE RELEASE HORMONE (LHRH)-LIKE IMMUNOREACTIVITY AMONG IN VITRO TRANSLATION PRODUCTS OF HYPOTHALAMIC mRNA

Richard Ivell
Dietmar Richter

Institut für Physiologische Chemie
Abteilung Zellbiochemie
Universität Hamburg
Hamburg, FRG

Rat hypothalamic mRNA was translated in the presence of [^{125}I]tyr-tRNA in a rabbit reticulocyte lysate system. Although no labelled products could be selected directly by anti-LHRH antisera, the introduction of a trypsinization step led to the identification of a specific product in the molecular weight range 21-28,000.

INTRODUCTION

The biosynthesis of a number of neuropeptide hormones has been elucidated in the last few years. Of those analysed (e.g. gastrin, adrenocorticotropin, met- and leu-enkephalin, somatostatin, arginine vasopressin, oxytocin, calcitonin, etc.) all are synthesized by ribosomal translation of mRNA via a longer precursor polypeptide, which is subsequently processed to give the functional end product. The biosynthesis of the releasing hormones, such as thyrotropin release hormone and luteinizing hormone release hormone (LHRH), have so far eluded such analysis, although the identity and structure of

BIOCHEMICAL AND CLINICAL ASPECTS OF
NEUROPEPTIDES: SYNTHESIS, PROCESSING,
AND GENE STRUCTURE

ISBN 0-12-417320-9

these peptides have been known for a number of years (Schally, 1978; Guillemin, 1978). LHRH is located at highest concentration in the median eminence of the hypothalamus (Palkovits et al., 1974), though debate is still active as to whether or not the hypothalamus is the site of LHRH synthesis (Carillo, 1980). It is also reported to be endogenous to the pancreas (Seppälä et al., 1979), the testis (Dutlow and Millar, 1981) and the placenta (Khodr and Siler-Khodr, 1980). In 1973, it was postulated that LHRH may be synthesized via an enzyme template system like some of the peptide antibiotics (Johansson et al., 1973); this work was not subsequently developed. Since then, reports have emerged of a large molecular weight polypeptide isolable from hypothalamus (Millar et al., 1977; Millar et al., 1981) or testis (Dutlow and Millar, 1981) that reacts specifically with anti-LHRH antisera. It should be pointed out, however, that other groups have also looked for such high molecular weight immunoreactive material and with one exception (Gautron et al., 1981) have failed to find it (e.g. Barnea and Porter, 1980).

Following our recent successes in identifying precursors to the neurophysins, arginine vasopressin, oxytocin and to somatostatin (Schmale and Richter, 1980; Land et al., 1982; Ivell and Richter, 1982) by in vitro translation of hypothalamic mRNA, we have applied these techniques to the quest for an LHRH mRNA-dependent precursor. Here we present evidence for the in vitro biosynthesis of a polypeptide of 21-28 kdal, which following trypsin treatment has been tentatively identified using specific antibodies as containing an LHRH-like sequence.

MATERIALS AND METHODS

Isolation of mRNA

Initial experiments were carried out using RNA isolated by sodium dodecylsulfate (SDS)/phenol/chloroform extraction of polysomes separated by sucrose density gradient centrifugation from bovine hypothalami (Richter et al., 1980). The hypothalami were so dissected as also to include the preoptic and paraventricular nuclei which have been suggested as possible sites of LHRH synthesis (Barry et al., 1974; Knigge et al., 1977). This RNA was then passed over oligo(dT)-cellulose columns to enrich the poly (A)- containing species (Richter et al., 1980). In one experiment, the poly(A)$^+$RNA was subsequently subjected to isokinetic sucrose density gradient centrifugation (Richter et al., 1980) to obtain fractions considerably enriched in mRNA of a given size. Later, we repeated experiments using poly(A)$^+$ RNA isolated by the guanidinium thiocyanate procedure (Chirgwin et al., 1979) from rat hypothalami.

Translation

mRNA isolated as above was translated in a rabbit reticulocyte lysate translation system (New England Nuclear) in all cases including an optimal quantity of exogenous mRNA, to give maximum total incorporation of radiolabel into synthesized polypeptides. The translation mixtures were supplemented with (^{3}H)leucine (Amersham), (^{35}S)methionine, (^{35}S)cysteine or (^{125}I)tyr-tRNA (all from New England Nuclear), of the highest available specific activity and included at 25-100 μCi, 50 μCi, 20 μCi and 30 μCi, respectively per 25 μl reaction mixture. The (^{3}H)leucine, in ethanol solution, and (^{125}I)tyr-tRNA were both concentrated prior to use by lyophilization, or ethanol precipitation in the presence of excess tRNA^{Phe}, respectively. Because neither methionine nor cysteine appear in the decapeptide LHRH sequence, and (^{3}H)-labelling provided insufficiently radioactive translation products, most experiments were carried out with (^{125}I)tyr-tRNA, which in our hands was incorporated into hot trichloroacetic acid - precipitable hypothalamic protein to the extent of up to 500,000 cpm per 25 μl translation assay.

Immunoprecipitation

The majority of antisera raised against the functional decapeptide LHRH absolutely require the presence of the C-terminal amide group for recognition of the antigen (unpublished, Schally et al., 1980). In a hypothetical precursor, this modification would not be present, the amide group probably deriving from a glycine residue C-terminal to the decapeptide (Suchanek and Kreil, 1977). Thus only antibodies can be used for immunoprecipitation of a potential precursor to LHRH which do not require the C-terminal amide for recognition. Such antibodies must also be able to recognize an LHRH molecule extended at both N- and C-termini. In our experiments we have used an antibody (§ 743), which appears to possess the required qualities. This antiserum has been well characterized (Schally et al., 1980) and we are particularly grateful to Dr. A. Arimura for generously donating it for our research. Significantly this is one of the antisera with which Millar and coworkers have identified their large molecule weight immunoreactive material (Millar et al., 1981).

Immediately following translation, the rabbit reticulocyte lysates (25-100 μl) were mixed with three volumes of either buffer A (10 mM $NaHPO_4$, pH 7.5, 1 mM ethylene diamine tetraacetic acid (EDTA), 1 % NP-40) or buffer B (as buffer A plus 1 % sodium deoxycholate, 0.3 % SDS), together with 2-8 μl Antagosan (Behring), unlabelled cysteine and methionine (where used as radiolabel) to 5mM and dithiothreittol to 2 mM final concentrations. IgG fractions were prepared from the antiserum either by 40 % (v/v) saturated ammonium sulfate precipitation (Richter et al., 1980) or by Protein A-Sepharose affinity chromatography (Goudswaard et al., 1978), and added to the above immunoprecipitation mix in quantities varying from 20 to 150 μg per tube.

After incubation overnight at 4°, possible immunoreactive products were isolated using either formaldehyde-fixed Staphylococcus aureus immunoabsorbant (Richter et al., 1980) or Protein A-Sepharose (Ivell and Richter, 1982), washing the immunoabsorbant several times in buffer C (0.5 M NaCl, 1 mM EDTA, 20 mM Tris·HCl, pH 7.6, 0.5 % NP-40), then buffer D (as buffer C, but with only 0.1 M NaCl), before eluting immune complexes in electrophoresis sample buffer (62.5 mM Tris·HCl, pH 6.8, 5 % SDS, 10 % glycerol, 4 % betamercaptoethanol) or 1 M acetic acid, respectively. The latter procedure, using Protein A-Sepharose,was found considerably to reduce the nonspecific background in the subsequent electropherograms. The acetic acid eluate was lyophilized before being dissolved in electrophoresis sample buffer. Samples were electrophoresed on either 10-20 % gradient or 15 % polyacrylamide/SDS gels, which were subsequently fluorographed (Bonner and Laskey, 1974) and laid on film for up to 3 months. In the case of ^{3}H-labelling, the films were pre-flashed (Laskey and Mills, 1977), and in the case of ^{125}I-labelling, intensifying screens (Du-Pont Cronex Lightning Plus) were also used (Laskey and Mills, 1977).

As criterion by which to judge the resulting autoradiograms, a result was considered positive if a band appeared in a gel track which did not appear in parallel tracks of (a) immune precipitates which had been supplemented with 20-100 μg of unlabelled LHRH (UCB Bioproducts, Brussels) or LHRH free acid (Peninsula, San Carlos, California) prior to the addition of the antibodies, (b) immune precipitates derived from translations where no hypothalamic mRNA has been included, and (c) immune precipitates made with pre-immune rabbit IgG instead of anti-LHRH.

Tryptic Digestion of Total Translation Products

Although large molecular weight LHRH-like material has apparently been identified without recourse to hydrolysis, Millar et al. (1977) claim that trypsinization can enhance the immunoreactivity up to 20-fold. Because of an adjacent proline residue in LHRH, the arginine at position 8 is not cleaved by trypsin. We therefore modified our procedures to include a trypsinization step in the analysis. Using (^{125}I)tyr-tRNA as radiolabel, rat hypothalamic mRNA was translated in a rabbit reticulocyte lysate system and the total translation from six 25 μl reaction mixtures electrophoresed on a 15 % polyacrylamide/SDS gel. The gel was dried immediately, just before the bromophenol blue front was eluted, and laid on film for 7 days. Guided by the resulting autoradiogram, the gel was cut into ten 1 cm horizontal stripes, each including proteins of a given size range. Proteins were eluted from these gel pieces by stirring the macerated gel for 4 hr at 37° in 10 ml of 1 % SDS containing lysozyme (Sigma, grade I) at 100 μg/ml. After centrifuging the gel debris (5000 g, 15 min, 10°) and washing it in a further 3 ml of 1 % SDS, the supernatants were combined and precipitated at -25° overnight by addition of 3 volumes of acetone. The resulting precipitates were dried under vacuum and redissolved in 0.8 ml of 0.1 M NH_4HCO_3, pH 8.0, to which 100 μg tosylamide phenylethyl chloromethyl ketone - treated trypsin (Worthington) was then added, and incubated at 37° overnight. The following morning a further 50 μg trypsin was added to each digest and incubation continued for additional 6 hours. At the end of this time 0.5 mg of soybean trypsin inhibitor (Sigma, Type 1-s) was added and the tubes left on ice for 30 min. The digests were then adjusted to 1 % in NP-40 and 0.5 mM in EDTA, divided into equal paired aliquots, to one of which was added 20 μg unlabelled $LHRH_{free\ acid}$, to the other 20 μg unlabelled arginine vasopressin (Calbiochem) as control, followed by the addition of 25 μg anti-$LHRH_{\#743}$ to all tubes. The immune reaction was allowed to proceed overnight at 4°, before isolating immune complexes using Protein A-Sepharose as described above. The acetic acid eluates were then counted directly in a gamma-scintillation counter. The treatment of the translations prior to electrophoresis with RNAase A (Boehringer, 8 μg/ml), although eliminating the large (^{125}I)tyr-tRNA band on the autoradiograms, appeared to have no effect on the final results. Control experiments were also carried out to determine the influence of the procedure on the antigen-antibody reaction. Thin-layer chromatography showed that trypsin had no influence on the LHRH decapeptide when incubated under the conditions of the experiment in the presence of carrier lysozyme. Radioimmunoassays showed that although trypsin alone clearly affects the immune reaction, the soybean trypsin inhibitor is fully able to prevent any negative effect to the enzyme under the conditions used in the above experiment. Parallel experiments were also made using total translation mixtures incubated in the absence of added hypothalamic mRNA.

RESULTS AND DISCUSSION

In the absence of trypsin treatment, we were unable reliably to identify any large molecular weight material having LHRH-like properties. In occasional experiments, apparently positive results were obtained, corresponding to a molecular weight range of 20 to 30 kdal, but the protein bands on the autoradiograms were barely distinct from the background even after long exposures and proved very difficult to repeat.

Somewhat better results were obtained after the introduction of a tryptic digestion step into the procedures as described in Materials and Methods. Total translation products were first separated by SDS/polyacrylamide gel electrophoresis, and proteins of a given size range were eluted from the gel and trypsinized before participating in the immune reaction with specific anti-LHRH antibodies. Figure 1C illustrates a result from this type of experiment. A significant peak of LHRH-like activity is evident, corresponding to a molecular weight of 21 to 28 kdal. This peak is not present in the parallel immune reactions where 20 µg of unlabelled LHRH was included prior to addition of the antibodies (Fig. 1D). Nor has LHRH-like material been detected when translations, made in the absence of added hypothalamic mRNA, were used as starting material (not shown). Attemps to analyse the LHRH-like peptides further have so far not proved possible because of the very small quantity of material available. Inspection of the autoradiograms in the region 21 - 28 kdal (Fig. 1B) indicates several small bands attributable to the hypothalamic mRNA. Whether one of these bands is an LHRH precursor requires further research. Interestingly, the apparent molecular weight of the LHRH-like material we have tentatively identified corresponds well with the approximately 26 kdal LHRH-polypeptide identified by Gautron et al. (1981) as well as the cell-free translated 28 k product described in Curtis and Fink, 1983.

Because we are working at the limits of resolution of our analytical system, these results must be interpreted cautiously. They give weight to the view that LHRH is probably synthesized via a longer precursor by translation of mRNA within the hypothalamus, supporting, therefore, the results of those groups who have identified large molecular weight LHRH-like material in hypothalamic extracts (Millar et al., 1981; Gautron et al., 1981). The antiserum we have been using apparently recognizes a tryptically digested precursor better than the original longer molecule, as had also been observed by Millar et al., 1977. By analogy with known sequences of other peptide hormone precursors, they predicted that the decapeptide LHRH is probably flanked by basic amino acids in the longer sequence, and that

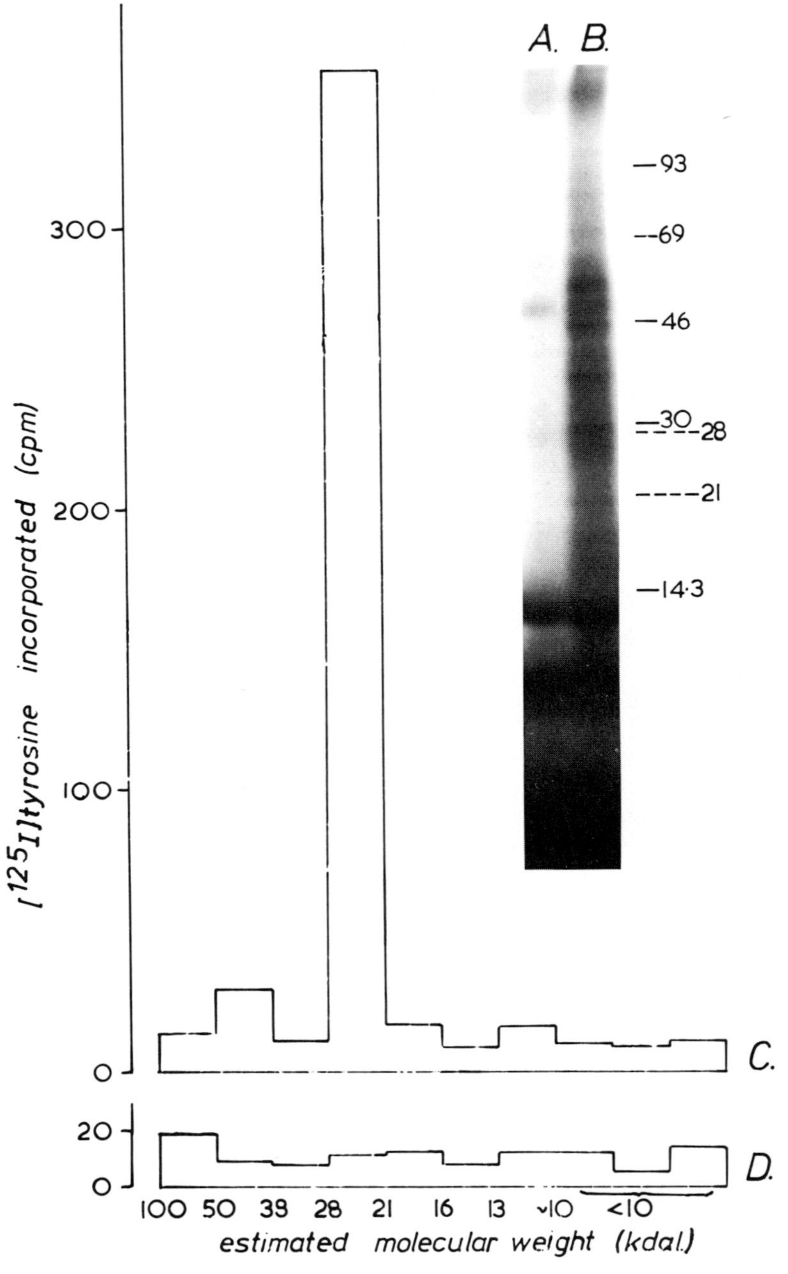

Figure 1. Incorporation of (^{125}I)tyrosine into LHRH-like immunoreactive polypeptides. A,B: fluorograms of 10-20 % gradient polyacrylamide/SDS gel tracks each containing 3 µl of a translation mix

tryptic cleavage would yield a peptide only slightly longer than LHRH and thus more accessible to the antibodies (Millar et al., 1981).

LHRH would appear to be synthesized in very small quantities. A comparison of our results on the in vitro biosynthesis of arginine vasopressin (Schmale and Richter, 1980; Land et al., 1982) and somatostatin (Ivell and Richter, 1982) by translation of hypothalamic mRNA suggests that the mRNA for the latter hormone is considerably less frequent than that for arginine vasopressin, corresponding to the relative quantities of these peptides determined in vivo [arginine vasopressin: 500 ng/mg rat hypothalamic protein (George and Jacobowitz, 1975), but very much more also of hypothalamic origin is stored in the posterior pituitary (Valtin et al., 1974); somatostatin: 500 ng/mg rat hypothalamic protein (Brownstein et al., 1975; Terry and Crowley, 1980)]. LHRH has been measured in vivo to be approximately one order of magnitude less frequent than somatostatin [LHRH: 30 ng/mg rat hypothalamic protein (Palkovits et al., 1974; Advis et al., 1980; Simpkins et al., 1980)], implying that its mRNA may also be correspondingly rarer. If this is the case, a new strategy would be necessary to elucidate LHRH biosynthesis in an adequate fashion. One hope is that a tumor cell line may be found that

---

supplemented (B) or not (A) with 0.6 µg rat hypothalamic mRNA per 25 µl reaction mixture, also containing 30 µCi (^{125}I)tyr-tRNA. After translation, reaction mixtures were incubated for 15 min at 27° with RNAase A (8 µg/ml) to digest the unincorporated (^{125}I)tyr-tRNA. Marker proteins used were phosphorylase B, 96 kdal; bovine serum albumin, 69 kdal; ovalbumin, 43 kdal; carbonic anhydrase, 30 kdal; lysozyme, 14.3 kdal. C,D: Histograms showing LHRH-like immunoreactivity in the tryptic digests of polypeptides eluted from a 15 % polyacrylamide/SDS gel, as described in Materials and Methods. The immune reactions were supplemented with 20 µg arginine vasopressin (C) or 20 µg $LHRH_{free\ acid}$ (D) prior to the addition of antibodies.

produces LHRH in large quantities. Alternatively, recombinant DNA technology may offer possibilities, although this approach would involve the screening of a large number of clones; If LHRH is synthesized in as low proportions as the corticotropin release factor whose precursor sequence was recently identified from a hypothalamic cDNA bank (Furutani et al., 1983) analogously more than 10^6 clones may have to be screened.

REFERENCES

Advis, J.P., McCann, S.M. and Negro-Vilar, A., *Endocrinology 107*, 892 (1980).

Barnea, A., Cho, G. and Porter, J.C., *Endocrinology 106*, 1980 (1980).

Barry, J., Dubois, M.P. and Carette, B., *Endocrinology 95*, 1416 (1974).

Bonner, W.M. and Laskey, R.A., *Eur. J. Biochem. 46*, 83 (1974).

Brownstein, M., Arimura, A., Sato, H., Schally, A.V. and Kizer, J.S., *Endocrinology 96*, 1456 (1975).

Carillo, A.J., *Life Sciences 27*, 179 (1980).

Chirgwin, J.M., Przybyla, A.E., MacDonald, R.J. and Rutter, W.J., *Biochemistry 18*, 5294 (1979).

Curtis, A. and Fink, G., *Endocrinology 112*, 390 (1983).

Dutlow, C.M. and Millar, R.P., *Biochem. Biophys. Res. Commun. 101*, 486 (1981).

Furutani, Y., Morimoto, Y., Shibahara, S., Noda, M., Takahashi, H., Hirose, T., Asai, M., Inayama, S., Hayashida, H., Miyata, T. and Numa, S., *Nature 301*, 537 (1983).

Gautron, J.P., Patton, E. and Kordon, C., *Mol. Cell. Endocrinology 24*, 1 (1981).

George, J.M. and Jacobowitz, D.M., *Brain Res. 93*, 363 (1975).

Goudswaard, J., Van der Donk, J.A., Noordzij, A., Van Dam, R.H. and Vaerman, J.P., *Scand. J. Imunol. 8*, 21 (1978).

Guillemin, R., *Science 202,* 390 (1978).

Ivell, R. and Richter, D., *Biochemistry 21,* 1204 (1982).

Johansson, K.N., Currie, B.L. and Folkers, K., *Biochem. Biophys. Res. Commun. 53,* 502 (1973).

Khodr, G.S. and Siler-Khodr, T.M., *Science 207,* 315 (1980).

Knigge, K.M., Hoffman, G., Scott, D.E. and Sladek, J.R., *Brain Res. 120,* 393 (1977).

Land, H., Schütz, G., Schmale, H. and Richter, D., *Nature 295,* 299 (1982).

Laskey, R.A. and Mills, A.D., *FEBS Lett. 82,* 314 (1977).

Millar, R.P., Aehnelt, C. and Rossier, G., *Biochem. Biophys. Res. Commun. 74,* 720 (1977).

Millar, R.P., Wegener, I. and Schally, A.V., *in* "Neuropeptides: Biochemical and Physiological Studies" (R.P. Millar, ed.), pp. 111. Churchill Livingstone, New York, (1981).

Palkovits, M., Arimura, A., Brownstein, M., Schally, A.V., and Saavedra, J.M., *Endocrinology, 95* 554 (1974).

Richter, D., Schmale, H., Ivell, R., and Schmidt, C., *in* "Biosynthesis Modification and Processing of Cellular and Viral Polyproteins", (G. Koch and D. Richter, eds.) pp. 43-66, Academic Press, New York (1980).

Schally, A.V., *Science 202,* 18 (1978).

Schally, A.V., Arimura, A., and Coy, D.H., *Vitamins and Hormones 38,* 258 (1980).

Schmale, H. and Richter, D., *FEBS Lett. 121,* 358 (1980).

Seppälä, M., Wahlstrom, T. and Leppäluoto, J., *Life Sciences 25,* 1489 (1979).

Simpkins, J.W., Kalra, P.S. and Kalra, S.P., *Endocrinology 107,* 573 (1980).

Suchanek, G. and Kreil, G., *Proc. Natl. Acad. Sci. USA 74,* 975 (1977).

Terry, L.C. and Crowley, W.R., *Endocrinology 107,* 1771 (1980).

Valtin, H., Stewart, J. and Sokol, H.W., *Handbook of Physiol. 7,* 131 (1974).

PART VII

Expression of Neuropeptide Genes

PARTIAL SEQUENCE OF THE RAT OXYTOCIN GENE

Richard Ivell
Dietmar Richter

Institut für Physiologische Chemie
Abteilung Zellbiochemie
Universität Hamburg
Hamburg, FRG

Cloned DNA encoding the gene for the oxytocin/neurophysin common precursor was isolated from a rat genomic DNA library and chemically sequenced. The elucidated nucleotide structure showed that the small gene (<1000 bases) was divided into three exons, corresponding respectively 5' to 3' to: exon A, the signal sequence, nonapeptide hormone and initial amino acids of the associated neurophysin; exon B, the central region of the neurophysin; exon C, the C-terminus of neurophysin plus one supernumerary arginine residue. The gene structure is closely comparable to that of the vasopressin gene, the two genes in fact sharing 143 nucleotides within the second exon with only a single base change, strongly suggestive of a mutual gene conversion.

INTRODUCTION

The nonapeptide hormone oxytocin is synthesized in the mammalian hypothalamus in the form of a longer precursor polypeptide which also includes the sequence of the neurophysin carrier protein associated with the hormone in neurosecretory granules of the neurohypophysis (Pickering, 1982).

BIOCHEMICAL AND CLINICAL ASPECTS OF
NEUROPEPTIDES: SYNTHESIS, PROCESSING,
AND GENE STRUCTURE

ISBN 0-12-417320-9

This common precursor structure was implicated from results of *in vivo* labelling experiments (Russell *et al.*, 1980) as well as from in vitro translation of hypothalamic mRNA in a reticulocyte lysate or *Xenopus laevis* oocytes (Schmale and Richter, 1980). Subsequently, the amino acid sequence of the bovine oxytocin precursor was enumerated from the cDNA prepared from bovine hypothalamic mRNA (Land et al., 1983). Parallel studies of vasopressin biosynthesis revealed that this closely related nonapeptide was derived from a similarly organized precursor molecule (Schmale and Richter, 1982). The recent elucidation of the rat vasopressin gene structure (see accompanying article) showed that the three functional entities of the vasopressin precursor - the signal sequence and nonapeptide, the neurophysin carrier, and a glycoprotein of unknown function - were located on three different exons. It was therefore of great interest to elaborate the oxytocin gene structure also from the rat, partly to provide information on the structure and evolution of the hormone precursor and partly to give specific DNA probes with which to begin detailed physiological investigations of oxytocin gene transcription and its regulation in the rat.

METHODS

A rat genomic DNA library (partial Sau3A cleaved fragments inserted in lambda vector EMBL-3, kindly provided by Dr Gerd Scherer, Heidelberg) was screened for plaques which reacted positively with radiolabelled exons A and B of the rat vasopressin gene, but which failed to give a positive signal with exon C of this gene. The studies of bovine cDNA had shown this exon to be the least homologous region of the two mRNAs. Of the plaques fitting this requirement and whose DNA was subjected to restriction analysis, two types were observed, representing the same gene sequence in the two possible insertion orientations. One of these was mapped in greater detail and a 2.4kb EcoRI restriction fragment subcloned into plasmid pUC8. This subclone included the complete oxytocin gene, whose sequence was determined by the method of Maxam and Gilbert (Fig. 1).

RESULTS AND DISCUSSION

The partial sequence includes the entire structural region of the gene, subdivided, just as in the vasopressin gene, into three exons. Exon A includes a 19 amino acid signal sequence whose N-terminal methionine corresponds to a position 67 nucleotides downstream from a modified Goldstein-Hogness box (CATAAA). The signal sequence appears to be immediately succeeded by the oxytocin nonapeptide. The hormone is separated from the N-terminal 9 amino acids of the rat oxytocin-associated neurophysin by the sequence Gly-Lys-Arg, representing respectively the probable source for the hormone amidation (Schmale and Richter, 1983) and the dibasic cleavage signal typical of other cellular polyproteins. There follows a short intron of ca. 220 nucleotides, identified by the consensus GT and AG bases at its 5' and 3' ends respectively. The second exon includes the sequence coding for the central 67 amino acids of the neurophysin and is followed by a very short intron of 86 nucleotides. Exon C codes for the remaining 17 amino acids of neurophysin, as it has been sequenced directly by Acher and coworkers (Chauvet et al., 1981). Our sequence indicates a single arginine residue at the C-terminus additional to the protein sequence determined by the French group. Otherwise, the two neurophysin sequences are in perfect agreement. It is conceivable that the supernumerary arginine might be cleaved off from the precursor during post-translational processing. The bovine oxytocin cDNA indicates an extra histidine in the same position. This would tend to corroborate the idea of a post-translational cleavage step to remove the extra amino acid. A stop codon follows without any additional protein moiety corresponding to the C-terminal glycopeptide included in the vasopressin precursor of both calf and rat. An obvious polyadenylation site is present in the rat oxytocin gene 60 nucleotides downstream from the stop codon.

Thus the rat oxytocin gene is small, comprising less than 1000 bases, with the structural elements distributed on three different exons. The exon-intron organization is remarkably similar to that of the rat vasopressin gene (Schmale et al., 1983).

GACAAGG CATAAA AAGGTCGGTC

MET ALA CYS
TGGCTGGAGAAACCATCACCGACGGTGGATCTCGGACTGAACACCAACGCCATGGCCTG

PRO SER LEU ALA CYS CYS LEU LEU GLY LEU LEU ALA LEU THR SER ALA CYS TYR ILE GLN
CCCCAGTCTCGCTTGCTGCCTGCTTGGCCTACTGGCTCTGACCTCCGCCTGCTACATCCA

ASN CYS PRO LEU GLY GLY LYS ARG ALA ALA LEU ASP LEU ASP MET ARG LYS
GAACTGCCCCTTGGGCGGCAAGAGGGCTGCGCTAGACTTGGATATGCGCAAG GTGAGTCT

CYS LEU PRO CYS GLY PRO GLY GLY LYS GLY ARG CYS PHE GLY PRO
CCCCGA/ /CCCCGCAG TGTCTTCCCTGCGGACCCGGCGGCAAAGGGCGCTGCTTCGGGCC

SER ILE CYS CYS ALA ASP GLU LEU GLY CYS PHE VAL GLY THR ALA GLU ALA LEU ARG CYS
GAGCATCTGCTGCGCGGACGAGCTGGGCTGCTTCGTGGGCACCGCCGAGGCGCTGCGCTG

GLN GLU GLU ASN TYR LEU PRO SER PRO CYS GLN SER GLY GLN LYS PRO CYS GLY SER GLY
CCAGGAGGAGAACTACCTGCCCTCGCCCTGCCAGTCTGGCCAGAAGCCTTGCGGAAGCGG

GLY ARG CYS ALA THR ALA GLY ILE CYS CYS SER PRO A
AGGCCGCTGCGCCACCGCGGGCATCTGCTGTAGCCCCGG GTGAGCAGGA/ /TCTCTTGCAG

```
SP GLY   CYS   ARG   THR   ASP   PRO   ALA   CYS   ASP   PRO   GLU   SER   ALA   PHE   SER   GLU   ARG   ***
A T G G C T G C C G C A C C G A C C C C G C C T G C G A C C C T G A G T C T G C C T T C T C C G A G C G C T G A G C C C

T G C T T T G T G A T G A T A C C T T T A G G G C G C T T C C T T C G T T C C C C A T G G C C A C T A C C A G A A A A A
```

Figure 1. Nucleotide sequence of the rat oxytocin gene. The relevant amino acid sequence for the corresponding precursor protein is indicated. Subclone pOλEco was made by inserting a 2.4kb restriction fragment of the isolated lambda clone into plasmid pUC8. Restriction fragments from this subclone were ^{32}P labelled either at their 5' ends with (^{32}P)γATP and T4 polynucleotide kinase or at their 3' ends using (^{32}P)cordecypin and terminal deoxynucleotidyl transferase. These fragments were sequenced by the method of Maxam and Gilbert, 1980, such that all illustrated sequence could be read at least once from each DNA strand. The modified Goldstein-Hogness sequence (CATAAA) is enclosed in a box; a stretch of 143 nucleotides, with a single interruption, common to both oxytocin and vasopressin genes from the rat is underlined.

More striking is a region of almost complete homology of both nucleotides and amino acids within exon B. In a stretch of 143 nucleotides there is only a single base change corresponding to the replacement of a leucine in the vasopressin precursor by valine (Chauvet et al., 1981). Since in the rat neither introns nor immediately adjacent exon sequences are involved, the homology points to a gene conversion event rather than alternative splicing as a possible explanation, with a point mutation having occured subsequently.

Comparison of the two nonapeptide gene sequences also shows, as anticipated, homology in exon A as well as exon B, indicative of the common origin of the two genes. Therefore, probes which will absolutely discriminate between the two gene transcripts can only be derived from the region of the third exon. It is hoped soon to be able to use such probes to investigate the regulation of hypothalamic nonapeptide synthesis. Recently, oxytocin has also been identified in extra-hypothalamic tissue, particularly the late corpus luteum (Wathes and Swann, 1982). Using oxytocin-specific DNA probes it should now be possible to determine whether these tissues are sites of hormone synthesis as well as to investigate the role of ectopic oxytocin in, for example, the luteal cycle.

REFERENCES

Chauvet, M.T., Chauvet, J. and Acher, R., *Biochem. biophys. Res. Commun. 103,* 595 (1981).

Land, H., Grez, M., Ruppert, S., Schmale, H., Rehbein, M., Richter, D. and Schütz, G., *Nature 302,* 342 (1983).

Maxam, A. M. and Gilbert, W., *Meth. Enzym. 65,* 499 (1980).

Pickering, B. T., *in* "Neurosecretion" (D. S. Farner and K. Lederis, eds.), pp. 415. Plenum Press, New York (1982).

Richter, D. and Schmale, H., *in* "Regulation of Gene Expression by Hormones" (K. W. McKerns, ed.), pp. 235. Plenum Publishing Corp., New York (1983).

Russell, J. T., Brownstein, M. J. and Gainer, H., *Endocrinology 107,* 1880 (1980).

Schmale, H. and Richter, D., *FEBS Lett. 121,* 358 (1980).

Schmale, H., Heinsohn, S. and Richter, D., *The EMBO Journal 2,* 763 (1983).

Wathes, D. C. and Swann, R. W., *Nature 297,* 225 (1982).

REGULATION OF GROWTH HORMONE GENE EXPRESSION

Michael G. Rosenfeld

University of California, San Diego
School of Medicine
La Jolla, California

Ronald M. Evans

The Salk Institute
Tumor Virology Laboratory
San Diego, California

Growth hormone is a pituitary polypeptide important for normal growth and development. Absent or decreased production is associated with a form of dwarfism. The growth hormone gene is an excellent model for studying complex control mechanism and its expression is regulated in the clonal rat pituitary cell lines G/C and GH_3 (Tashjian et al, 1968; Bancroft et al., 1969). In these cells it has been established that growth hormone biosynthesis is induced by glucocorticoids and by thyroid hormone and this induction is apparently the consequence of hormonal stimulation of growth hormone mRNA levels (Tushinski, 1977; Martial et al., 1977; Samuels et al., 1977; Seeburg et al., 1977; Harpold et al., 1979). The regulation of specific gene expression by glucocorticoids and related steroid hormones is postulated to be mediated by specific intracellular receptors which bind the hormone, undergo allosteric modification, and become associated with nuclear chromatin complexes (Yamamoto and Alberts, 1976; Samuels and Shapiro, 1967). Thyroid hormone is postulated to interact with nuclear thyroid hormone receptors (Samuels and Shapiro, 1976). The cloning of growth hormone cDNA in a number of laboratories, and the isolation of the growth hormone genomic clone (Cooke and Baxter, 1981; Page et al., 1981; Chien and Thompson, 1980; Gubbins et al., 1980; Doehmer et al., 1982) permitted more detailed analysis of the biochemical reactions modulating rat growth hormone gene expression. Analysis of the effects of the synthetic glucocorticoid, dexamethasone, or the accumulation of pulse-labeled growth hormone mRNA by increasing its rate of synthesis suggests that the primary site of action is

BIOCHEMICAL AND CLINICAL ASPECTS OF
NEUROPEPTIDES: SYNTHESIS, PROCESSING,
AND GENE STRUCTURE

ISBN 0-12-417320-9

nuclear (Evans et al., 1982). A series of studies which tentatively identify the growth hormone mRNA nuclear precursors (Harpold et al., 1979; Maurer et al., 1981; Dobner et al., 1981; Evans et al., 1982) permitted a demonstration that dexamethasone promotes an increase in growth hormone nuclear RNA precursors in proportion to its induction of mRNA levels (Doebner et al., 1981; Evans et al., 1982).

Transcription rates of the growth hormone gene were assessed by quantitation under DNA excess hybridization conditions of the number of nascent transcripts labeled with [^{32}P]-UTP following digestion in isolated nuclei (Evans et al., 1982), demonstrating a 3- to 4-fold increase in the transcription rate of the rGH gene in GC cells following addition of dexamethasone. Addition of dexamethasone has no effect on the transcription of at least two other specific gene products and actually decreased prolactin gene transcription. Induction is apparently maximal by 1 hour and persists at least 15 hours Fig. 1). In order to address the question of whether thyroid hormone and glucocorticoids can independently regulate growth hormone gene transcription, hormonal effects were evaluated in cells grown in serum-free conditions. The results presented in Fig. 2 establish that within 4 hours of its addition to cells in serum-free medium, either dexamethasone or T_3 can act independently to transcriptionally regulate the growth hormone gene. Simultaneous addition of both hormones results in a transcriptional stimulation greater than that produced by either agent alone.

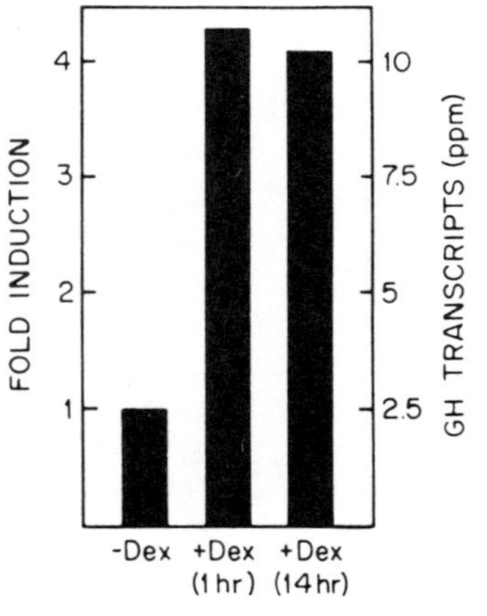

FIGURE 1. *Effects at 1 and 14 hours following addition of dexamethasone to G/C cell cultures on the transcription of the hormone gene.*

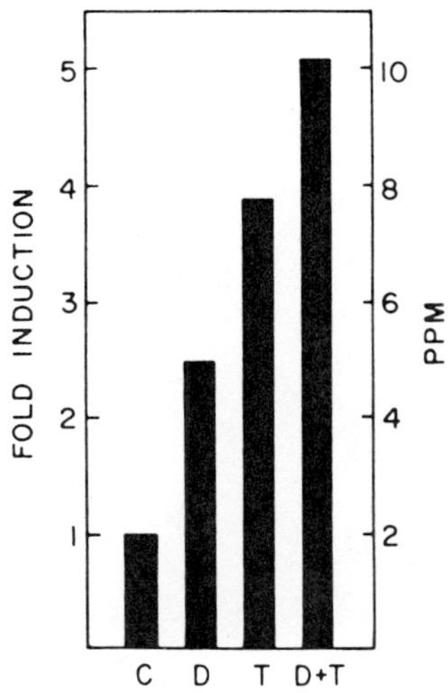

FIGURE 2. *Effects of thyroid hormone [T3 10^{-7}(T)] and dexamethasone (dexamethasone 10^{-6} M) on G/C cells under conditions of serum-free culture. Transcriptional rates were quantitated 5 h after hormonal addition (Evans et al., 1982).*

The physiological relevance of this regulation was evaluated by measuring the effects of glucocorticoids on the growth hormone gene expression in vivo revealed a 4-fold induction of GH gene transcription virtually identical to that observed with the G/C cell cultures in response to dexamethasone (Fig. 3).

The rat growth hormone gene contains determinants which transfer glucocorticoid regulation. Because glucocorticoids regulate growth hormone gene expression by exerting transcriptional regulation, it is critical to establish whether this regulation can be transferred by the growth hormone gene itself. The ability to introduce purified eukaryotic genes into heterologous eukaryotic cells provides a powerful method for both selecting genes and for studying their regulation in a new cellular environment (Wigler et al., 1979; Pellicer et al., 1980; Buetti and Diggelman, 1981; Lee et al., 1980; Kurtz 1981; Lai et al., 1980. The GH gene was introduced into mouse 3T3 cells using a plasmic vector which contains the entire genome of M loney mouse sarcoma virus (Mo-MSV) (Fig. 4), in order to assess the functional importance of sequences involved in regulation of the growth hormone gene. The availability of the molecularly cloned MSV DNA to transform mouse fibroblasts was used as a dominant selective marker to construct mouse cell lines containing the rGH gene. Analyses of several lines of transformed cells suggested that most of the GH-related sequences had become chromosomally integrated and

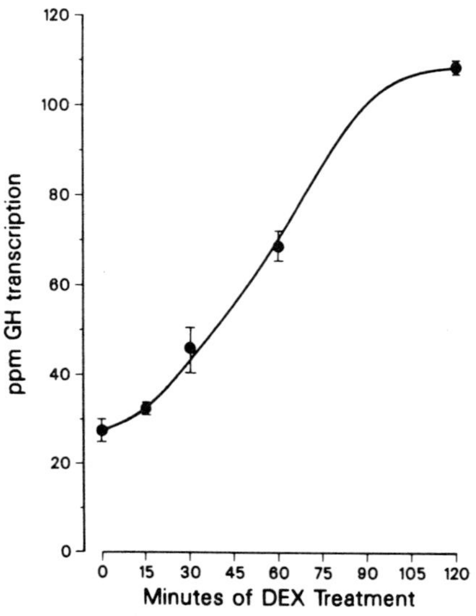

Figure 3. *Effect of dexamethasone administration in vivo to chronically adrenalectomized rats on growth hormone gene transcription in pituitary somatotrophs, as quantitated by analysis of nascent RNA transcripts (Evans et al., 1982).*

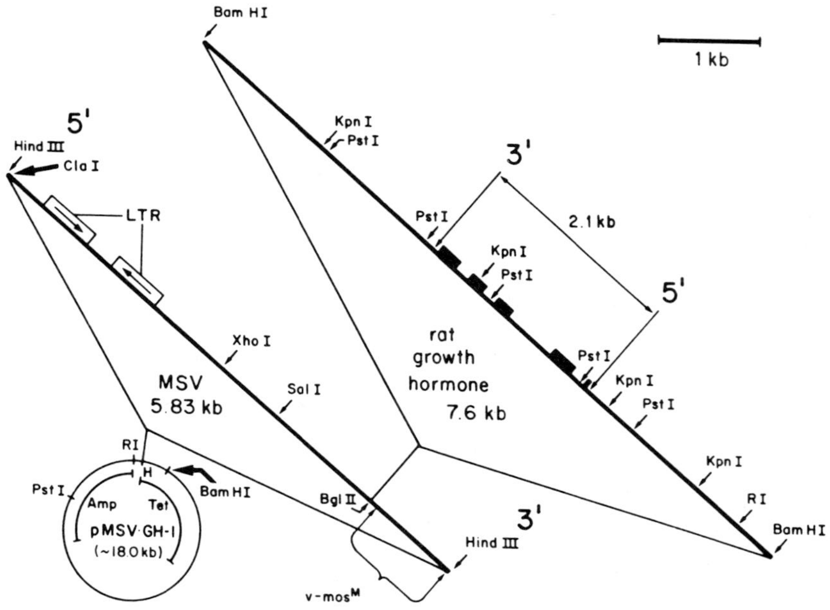

FIGURE 4. *Construction of the rat growth hormone Mo-MSV vector used in transfection experiments (Doehmer et al., 1982)*

the restriction endonuclease pattern of reactive fragments remained stable for successive generations, with independent transformants containing between 3 and 20 gene copies (Doehmer et al., 1982). These cells synthesized an mRNA approximately 250 nucleotides larger than rat pituitary GH mRNA, but which encoded the authentic growth hormone peptide, and secreted growth hormone (Doehmer et al., 1982). The potential effects of the synthetic glucocorticoid, dexamethasone, on GH production was examined (Fig. 5) because the 3T3 cells do contain glucocorticoid receptors. Addition of dexamethasone increased growth hormone production approximately 2.5-fold above that produced in unstimulated cells, quantitatively similar to the effects of dexamethasone on GC cell cultures. Similar transfer of regulation has been observed in the case of the human growth hormone gene (Robins et al., 1982). The ability to obtain regulated expression of the transfected rGH gene in heterologous cells suggests that hormonal responsiveness is intrinsic to the structure of the gene.

Expression of the transfected growth hormone gene in Vivo. The ability of the growth hormone gene to be expressed and regulated following transfection in heterologous cells suggested the possibility that introducing the cloned gene into the germ lines of heterologous species could provide further insight into eukaryotic gene regulation and development. The feasibility of introducing foreign DNA into the mammalian genome by microinjection of DNA molecules of interest into the pronuclei of fertilized eggs followed by insertion of the eggs into the reproductive tract of foster mothers has been established (Gordon et al., 1980; Brinster et al., 1981; Brinster et al., 1981; Wagner et al., 1981, 1982; Constantini and Lacy, 1981; Harbers et al., 1981; Rusconi and Schaffner, 1981). The growth hormone gene whose product has profound developmental effects provides a particularly convenient and biologically potent molecule to model the introduction of genes encoding secreted regulatory peptides into the germ line of animals. In a series of collaborative studies, such an approach was initiated by construction of a hybrid gene containing the mouse metallothionein promoter fused to the rat growth hormone gene (Fig. 6, Palmiter et al., 1983); this hybrid gene is reffered to as MGH. Palmiter et al. (1980) have reported that the metallothionein gene promoter did function in transgenic mice when fused to the structural portion of the herpes TK gene. Injection of the hybrid gene construction (MGH) into fertilized mice eggs by Dr. Brinster and his colleagues generated mice developing from these injected eggs which were significantly larger than their litter mates (up to 1.8-fold) and contained integrated copies of MGH in their genome.

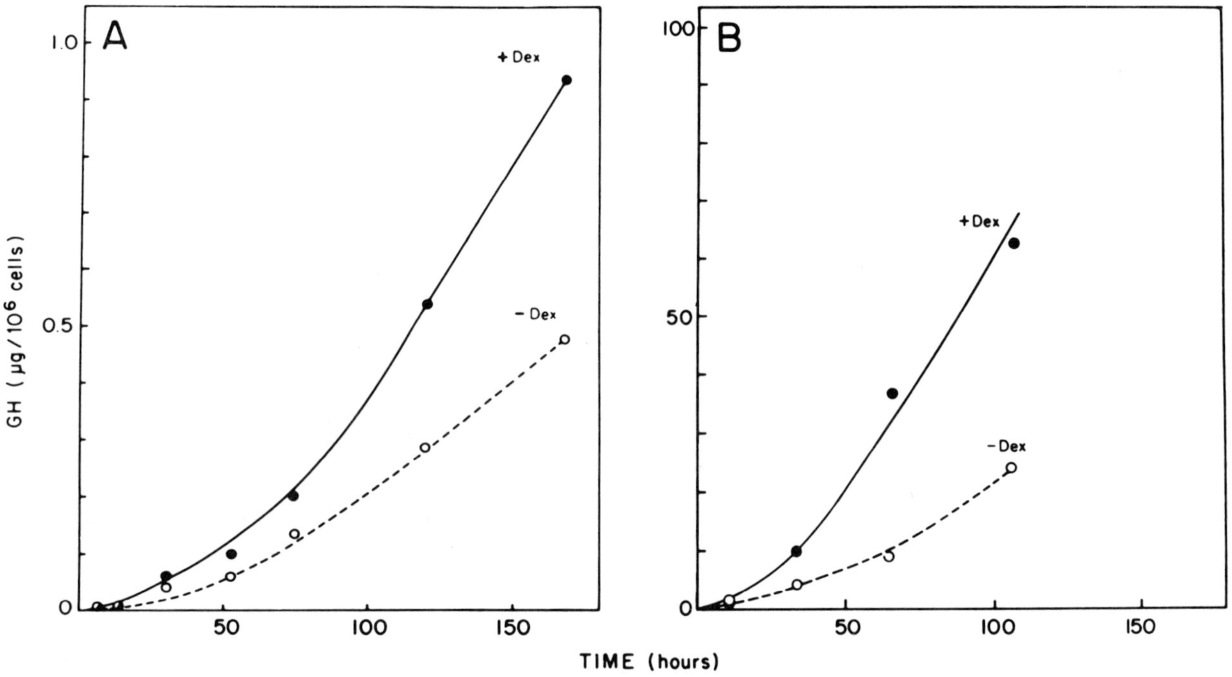

FIGURE 5. *Effect of dexamethasone on production of GH in cells transformed with pMSV-GH1 (A) and GC cells (B) (1 X 10^7/140-mm dish) in 25 ml of synthetic medium without serum were incubated with and without 5 µM dexamethasone.*

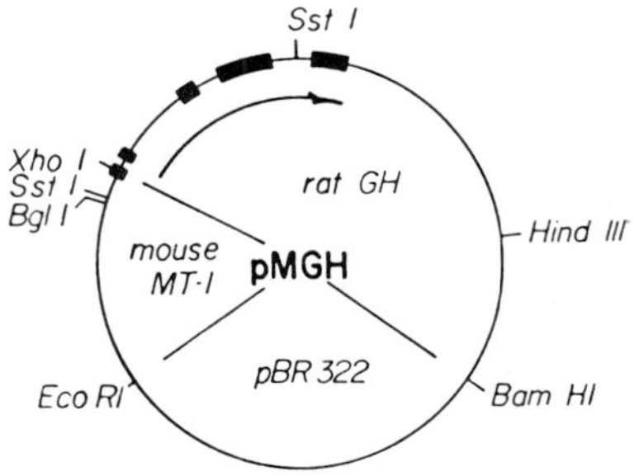

FIGURE 6. *Schematic diagram of the construction of fused genes containing a mouse metallothionein 1 (MT-1) promoter region and the rat growth hormone gene. The fused gene contains 68 bases of the MT-1 promoter and the entire rat growth hormone gene beginning at nucleotide 7 of the coding sequence.*

Evaluation of MGH gene expression in these so-called transgenic animals was initiated by analysis of its activity in liver, which revealed significant levels of reactive RNA. The level of MGH mRNA expression in the liver of the largest animals was unexpectedly high (800-3000 copies/cell) and correlated with the growth of the transgenic mice. The authenticity of the hybridized RNA as MGH RNA was confirmed by RNA blotting and S_1 nuclease analyses. The correlation between MGH mRNA levels and growth of the mice suggests that expression of the MGH gene accounts for the observed biological consequences and predict that circulating levels of GH should be elevated. The plasma GH values for four of the largest transgenic mice are 100 to 800 times greater than levels in control litter mates (Fig. 7). One mouse had 112 μg/ml of GH in its serum (3 orders of magnitude above physiological levels). Analysis of offspring of these transgenic mice revealed that the MGH gene was present and these animals were consistently large, with elevated plasma growth hormone levels. Furthermore, the introduced gene exhibited a tissue-specific pattern of expression; i.e., mRNA levels in liver, testes and gut exceed by >10-fold that in kidney and brain, and virtually no MGH mRNA is detectable in heart. The pattern of mRNA distribution is similar, but not identical, to that of the metallothionein gene product.

These data strongly suggest that the altered phenotype of these mice is a direct result of the integration and expression of the metallothionein-growth hormone fusion gene. In addition to the potential practical applications of this approach, it provides a tool to study the molecular basis of tissue-specific gene expression. We have initiated studies

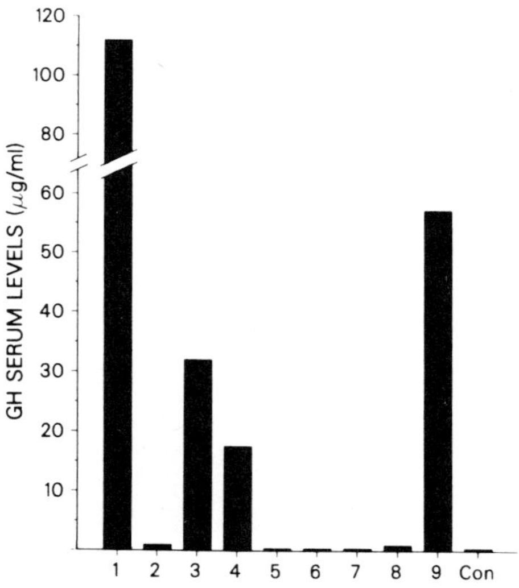

FIGURE 7. *Blood GH levels in transgenic mice. Animals 1, 3, and 9 represented the three animals showing most accelerated growth and highest content of GH mRNA in their livers.*

designed to analyze the tissue specificity of MGH gene expression. The tissue specificity of MGH gene expression is observed, compatible with the postulate that expression of the introduced gene is determined, at least in part, by sequences present in the introduced gene. Clearly, more knowledge is required with respect to the determinants of tissue-specific gene expression, efficiency of replication, and heritability to deduce how to optimally design hybrid genes for such a purpose.

REFERENCES

Bancroft, F. C., Levine, L., and Tashjian, A. H., Jr. *Proc. Natl. Acad. Sci. U.S.A. 78*, 4867 (1981).

Brinster, R. L., Chen, H. Y., Trumbauer, M., Senear, A. W., Warren, R., and Palmiter, R. D. *Cell 27*, 223 (1981).

Buetti, E., and Diggleman, H. *Cell 23*, 335 (1981).

Chien, Y. H., and Thompson, E. B. *Proc. Natl. Acad. Sci. U.S.A.* 77, 4583 (1980).

Constantini, F., and Lacy, E. *Nature (London) 295*, 92 (1981).

Cooke, N. E., and Baxter, J. D. *Nature (London) 297*, 603 (1982).

Dobner, P. R., Kawasaki, E. S., Yu, L. Y., and Bancroft, F. C. *Proc. Natl. Acad. Sci. U.S.A. 78*, 2230 (1982).

Doehmer, J., Barinaga, M., Vale, W., Rosenfeld, M. G., Verma, I. M., and Evans, R. M. *Proc. Natl. Acad. Sci. U.S.A. 79,* 2268 (1982).

Evans, R. M., Birnberg, N. C., and Rosenfeld, M. G. *Proc. Natl. Acad. Sci. U.S.A.*

Gordon, J. W., Scangos, G. A., Plotkin, D. J., Barbosa, J. A., and Ruddle, F. *Proc. Natl. Acad. Sci. U.S.A. 77,* 7380 (1980).

Gubbins, P., Lai, C.-J., Dhar, R., and Khoury, G. *J. Biol. Chem. 235,* 8655 (1980).

Harbers, K., Johner, J. D., and Jaenisch, R. *Nature (London) 292,* 540 (1981).

Harpold, M. M., Dobner, P. R., Evans, R., Bancroft, F. C., and Darnell, J. D. *Nucleic Acids Res. 6,* 3133 (1979).

Kurtz, D. T. *Nature (London) 291,* 629 (1981).

Lai, E. C., Woo, S. L. C., Bordelon-Riser, M. E., Fraser, T. H., and O'Malley, B. W. *Proc. Natl. Acad. Sci. U.S.A. 77,* 244 (1980).

Lee, F., Mulligan, R., and Ringold, G. *Nature (London) 294,* 228 (1981).

Martial, J. A., Baxter, J. D., Goodman, H. M., and Seeburg, P. H. *Proc. Natl. Acad. Sci. U.S.A. 74,* 1816 (1977).

Maurer, R. A., Gubbins, E. J., Erwin, C. R., and Donelson, J. E. *J. Biol. Chem. 255,* 2243 (1981).

Page, G., Smith, S., and Goodman, H. *Nucl. Acids Res. 9,* 2087 (1981).

Palmiter, R. D., Chen, H. Y., and Brinster, R. L. *Cell 29,* 701 (1982).

Palmiter, R. D., Brinster, R. L., Hammer, R. E., Trumbauer, M. E., Rosenfeld, M. G., Birnberg, N. C., and Evans, R. M. *Nature (London) 300,* 282 (1982).

Robins, D. M., Paek, I., Seeburg, P. N., and Axel, R. *Cell 29,* 623 (1982).

Rusconi, S., and Schaffner, W. *Proc. Natl. Acad. Sci. U.S.A. 78,* 5051 (1981).

Samuels, H. H., and Shapiro, L. E. *In* "Receptors and Hormone Action" (L. Birnbaumer and B. N. O'Malley, eds.), Vol. 13, pp. 35-75. Academic Press, New York.

Shapiro, L. E., Samuels, H. H., and Yaffee, B. M. *Proc. Natl. Acad. Sci. U.S.A. 75,* 45 (1978).

Tashjian, A. H., Jr., Yasumura, Y., Levine, L., Sato, G. H., and Parker, M. L. *Endocrinology 82,* 342 (1968).

Tushinski, R. J., Sussman, P. M., Yu, L. Y., and Bancroft, F. C. *Proc. Natl. Acad. Sci. U.S.A. 74,* 2357 (1977).

Wagner, E., Stewart, T., and Mintz, B. *Proc. Natl. Acad. Sci. U.S.A. 78,* 5016 (1981).

Wagner, T. E., Hoppe, P. C., Jollick, J. D., Scholl, D. R., Hodinka, R. L., and Gault, J. B. *Proc. Natl. Acad. Sci. U.S.A. 78,* 6376 (1981).

Yamamoto, K. R., and Alberts, B. M. *Ann. Rev. Biochem. 45,* 721 (1976).

EPIDERMAL GROWTH FACTOR mRNA CODES FOR A POTENTIAL POLYPROTEIN PRECURSOR

Axel Ullrich
Alane Gray
Thomas J. Dull

Department of Molecular Biology
Genentech, Inc.
South San Francisco, California

Epidermal growth factor (EGF) is coded for by a 4.75 kb messenger RNA in the male mouse submaxillary gland. The 53 amino acid long polypeptide sequence is present within a translational reading frame with the potential to code for an 1217 amino acid polypeptide. The presumptive precursor polypeptide contains 8 units of sequence homology to the EGF peptide. These homologous regions potentially represent additional functional segments within a polypeptide precursor.

I. INTRODUCTION

Mouse epidermal growth factor is one of the best characterized polypeptides known to stimulate proliferation of mammalian cells in culture (see Carpenter and Cohen, 1979 for review). It was first discovered in mouse submaxillary gland (MSG) extracts, as an activity which stimulated early eyelid opening when injected into newborn mice (Cohen, 1959, 1962). In addition to this profound effect on the differentiation of a specific cell type *in vivo*, EGF has been shown to be a potent mitogenic factor for a variety of cultured cells, of both ectodermal and mesodermal origin (see Carpenter and Cohen, 1979 for review).

BIOCHEMICAL AND CLINICAL ASPECTS OF NEUROPEPTIDES: SYNTHESIS, PROCESSING, AND GENE STRUCTURE

ISBN 0-12-417320-9

EGF is a 53 amino acid polypeptide (Taylor et al., 1972) containing 6 cysteine residues (Savage et al., 1972), which are thought to form 3 intrachain disulfide bonds (Savage et al., 1973). Urogastrone, a polypeptide bearing anti-gastric secretory activity isolated from human urine (Gregory, 1975), is presumably synthesized in submandibular and Brunner's glands and shares extensive sequence homology with EGF (70 percent) and may represent the human EGF equivalent.

EGF can be isolated from the adult male MSG. The level in the MSG appears to be regulated by androgens, since injection of testosterone into female or castrated male mice can increase the content of EGF from their significantly lower basal levels (Carpenter and Cohen, 1978).

EGF is obtained from the male MSG as a high molecular weight complex, comprised of 2 molecules EGF bound to 2 molecules of an EGF binding protein (Taylor et al., 1970; Taylor et al., 1974a). The EGF-specific binding protein is an arginine esteropeptidase (Taylor et al., 1974b) that appears to be involved in the processing of an EGF biosynthetic precursor (Frey et al., 1979).

In this report we present the sequence of a mouse EGF cDNA clone, which suggests that EGF is synthesized as a large protein precursor of 1217 amino acid residues. Our data indicate that the discrepancy between EGF levels in male and female MSGs is due to different EGF mRNA levels in these tissues, and suggest that the EGF-binding protein may have an unusual substrate specificity relative to other previously described, polypeptide hormone processing enzymes. The presumptive precursor polypeptide contains 8 regions with sequence homology to EGF. These may represent additional processing products with so far unknown functions.

II. METHODS

Procedures employed in this work have been described elsewhere in detail (Southern, 1975; Lehrach et al., 1977; Maxam and Gilbert, 1977; Wallace et al., 1981; Maniatis et al., 1982; Gray et al., 1983).

III. RESULTS AND DISCUSSION

A. Identification and Cloning of an EGF-specific cDNA fragment

Our cDNA cloning approach made use of the known amino acid sequence of mouse EGF (Savage et al., 1972 and Fig. 1) and employed male mouse submaxillary gland as the mRNA source. The carboxyl terminal region of EGF contains an amino acid sequence (TrpTrpGluLeu) which can only be coded for by four possible nucleotide sequences (Fig. 1). Thus four specific oligonucleotides, each eleven nucleotides in length, were synthesized to complement all possible nucleotide sequences which could code for this amino acid sequence. Their sequences are shown in Fig. 1; only the first two nucleotide positions of the leucine codon were included.

The following experiments were carried out to identify an EGF-specific cDNA fragment. The pooled synthetic oligonucleotides were 5'-end labelled by reaction with T4 polynucleotide kinase and γ-^{32}P-ATP and then used to prime cDNA synthesis on mRNA preparations from male and female mouse submaxillary glands. The male MSG contains significantly more EGF protein than the female MSG (Carpenter and Cohen, 1979); assuming that this was due to differences in mRNA levels, we used comparison of male and female MSG cDNA in an attempt to facilitate the identification of EGF cDNA. Acrylamide gel analysis showed that our oligonucleotides primed cDNA synthesis from a large variety of templates

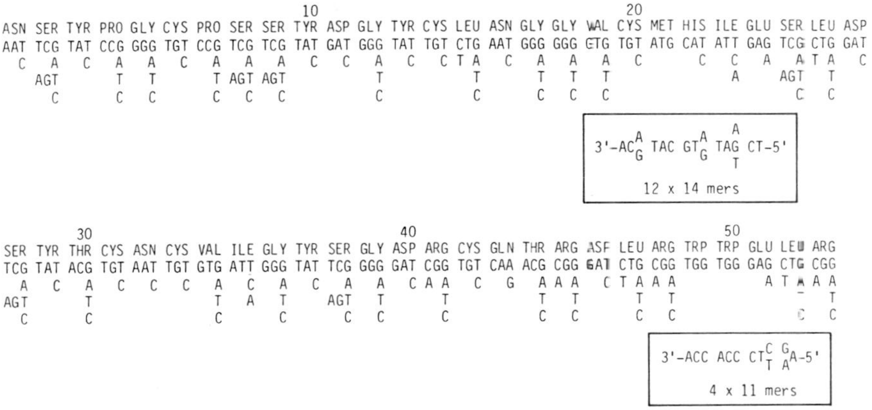

FIGURE 1. Mouse EGF amino acid sequence and potential nucleotide sequences. Oligonucleotide pools used for cDNA priming and screening are shown in boxes.

present in male and female MSG mRNA, and no visible pattern difference between male and female double-stranded (ds) cDNA could be detected (Fig. 2).

The fact that many cDNA species were synthesized indicated that several mRNAs of varying abundance contained sequences complementary to our oligonucleotide primers. Since no difference was detected between the male and female MSG mRNA preparations, it was possible that either the EGF-mRNA was a minor component of the male mRNA and not discernible over the more abundant background, or that the

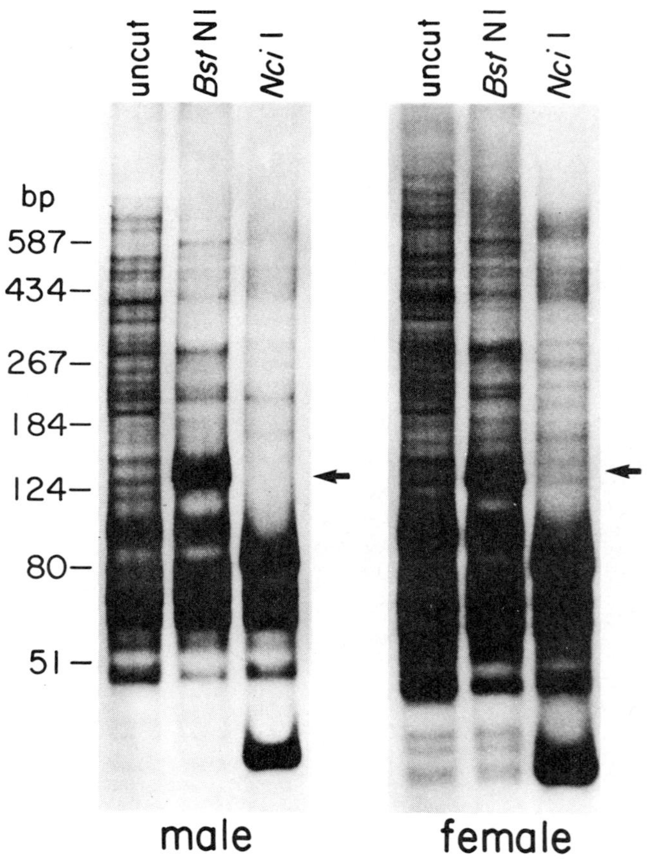

FIGURE 2. Comparison of ds cDNA synthesized by priming male or female MSG polyadenylated RNA with a pool of 4 synthetic undecamers. The arrow indicates the position of the predominant ~145 bp restriction digestion product which was eluted from the dried gel for Maxam-Gilbert sequencing.

difference in protein levels between the male and female glands was not due to differences in mRNA levels, as we had assumed. Restriction enzyme cleavage analysis was undertaken to resolve this point.

On the basis of the EGF amino acid sequence we determined all possible restriction fragment lengths which could be generated by priming cDNA synthesis with our specific oligonucleotides and digestion of the resulting ds cDNA with the restriction endonucleases BstNI and NciI. These particular enzymes were chosen because the amino terminal region of EGF contains a Pro-Gly dipeptide which is coded for by CCNGGN (N=A,T,G or C) and represents a restriction site for either BstNI (CC$^{T}_{A}$GG) or NciI (CC$^{C}_{G}$GG). Given this known cleavage site and the distance between the cleavage site and the primer, it could be calculated that the maximum size of a EGF cDNA fragment generated by BstNI or NciI cleavage would be 144 bp. The generation of shorter fragments (BstNI: 23 bp, 43 bp, 79 bp 131 bp, 134 bp, 144 bp; NciI: 23 bp, 72 bp, 43 bp, 143 bp, 144 bp, 145 bp) was predicted to be of a lower but finite probability. As can be seen in Fig. 2, fragments of the predicted possible size were generated but were not specific for the male MSG cDNA. To investigate the possibility of the presence of EGF mRNA in the female gland we determined the nucleotide sequence of the common ~145 bp BstNI fragment. The sequence did not contain any EGF coding information (data not shown). It was still possible that a minor restriction fragment was masked in the male cDNA by the non-EGF cDNA fragment. However, DNA sequence analysis of the common ~145 bp fragment revealed the presence of a HaeIII cleavage site, which computer analysis demonstrated could not be present in EGF coding sequences. We therefore repeated our BstNI/NciI cleavage analysis of male and female primed cDNA after digestion with HaeIII to remove the common ~145 bp fragment. In order to improve resolution of the analysis we electrophoresed primed, HaeIII digested ds cDNA first on an acrylamide gel, cut the gel into 0.5 cm slices, and eluted each individual gel slice. Each cDNA fraction was then digested with BstNI or NciI and analyzed on another acrylamide gel. Autoradiography of the dried gel showed that a ~145 bp HaeIII/BstNI fragment was indeed present in the cleaved male cDNA and not in the female (Fig. 3). The fractions that yielded the ~145 bp male specific fragment covered a size range of 150-200 bp.

Our cloning approach included HaeIII digestion as a sequence enrichment step in addition to specific oligonucleotide priming. Oligonucleotide primed-male ds cDNA was digested with HaeIII and subsequently electrophoresed on a 6 percent acrylamide gel. Four size cuts covering a 120-300 bp

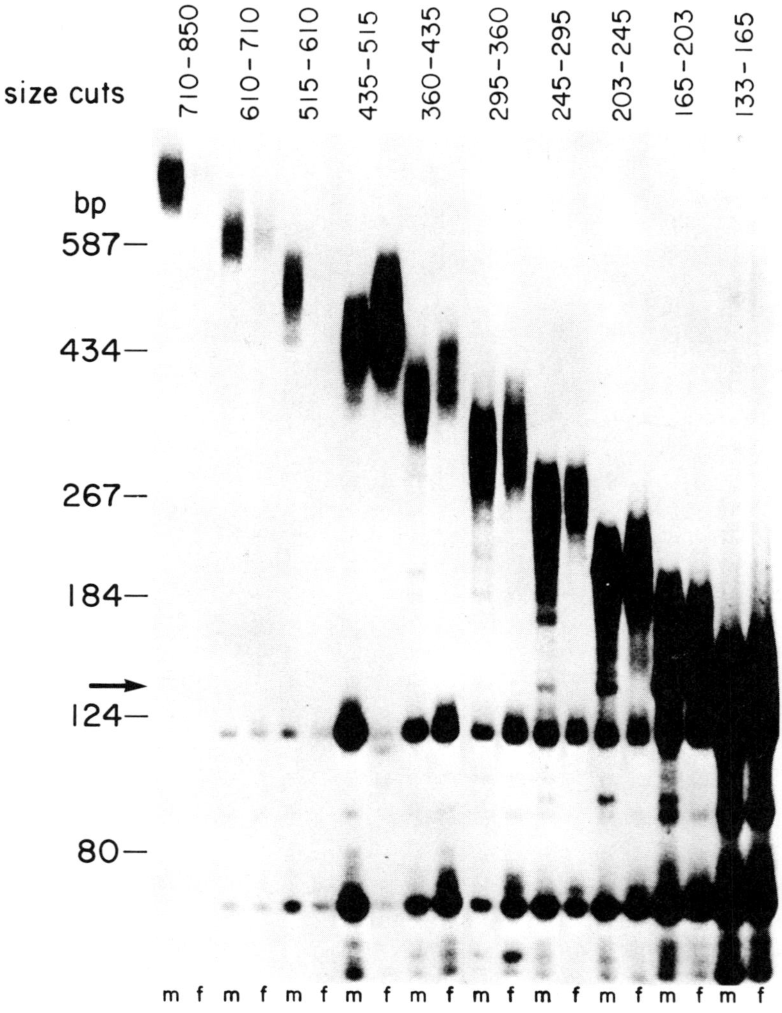

FIGURE 3. Comparison of restriction endonuclease digestion products of various size fractions of HaeIII-digested male and female MSG ds cDNA. ^{32}P-labelled pBR322 HaeIII fragments were used as size markers. Only the BstNI digestion is shown. The size cut range (bp) for each pair of lanes is indicated at the top. m = male MSG polyA^{+}-RNA used as template; f = female. The arrow points at the ~145 bp male specific BstNI fragment. Only those fragments which contain the radiolabeled primer can be detected on this autoradiograph.

range were taken and cDNA was electrophoretically eluted. We omitted S1 nuclease treatment and cloned the cDNA directly into the PstI site of pBR322 using the standard dC/dG tailing procedure (Wickens et al., 1978), hoping to tail from a probable HaeIII site upstream from the sequence coding for EGF. 600 clones from each fraction were screened with the oligonucleotides originally used for cDNA priming (Wallace et al., 1981). Six percent of the clones hybridized under stringent conditions and were rescreened with another pool of twelve, 14 nucleotide-long oligonucleotides which would complement an amino acid sequence upstream from the TrpTrpGluLeu sequence (see Fig. 1). In addition, ^{32}P-labelled single-stranded (ss) cDNA primed with the pool of four, 11 nucleotide-long oligonucleotides on male and female MSG mRNA was used as hybridization probe in parallel experiments. Four clones were identified which hybridized with both of the synthetic DNA probes and the male ss cDNA but not the female. Restriction enzyme and partial DNA sequence analysis using the enzymatic method by Sanger (1977) confirmed that all the clones were identical and contained EGF coding sequences. Clone pmEGF-C7 was sequenced in its entirety by the Maxam-Gilbert method (Maxam and Gilbert, 1977). Its 171 bp DNA insert begins at the 3' end with the primer sequence TGGTGGGAGCT, and ends at the 5' end with a HaeIII site as planned.

B. A 4.75 kb mRNA encodes mouse EGF

Determination of the size of the male MSG EGF mRNA was critical to our isolation of a full-length cDNA, and we were interested to test if these sequences were detectable in mRNA isolated from the female gland. Male and female MSG mRNA was subjected to electrophoresis in a formaldehyde containing agarose gel (Lehrach et al., 1977), and analyzed by Northern blotting using a radiolabelled (Taylor et al., 1976) 171 bp, HaeIII-SstI restriction fragment of the clone pmEGF-C7 as probe (see Fig. 4). As expected, there was no detectable hybridization with the female MSG mRNA, however, to our surprise, our EGF probe hybridized to an RNA species of about 4750 nucleotides in length (Fig. 4). Considering that the EGF protein requires only 159 nucleotides to encode it, the existence of an mRNA 30 times this size was a remarkable finding.

In order to isolate cDNA clones which contained the entire sequence for EGF mRNA we used specific priming with DNA restriction fragments and synthetic DNA as well as oligo dT priming. The nucleotide sequence of seven overlapping

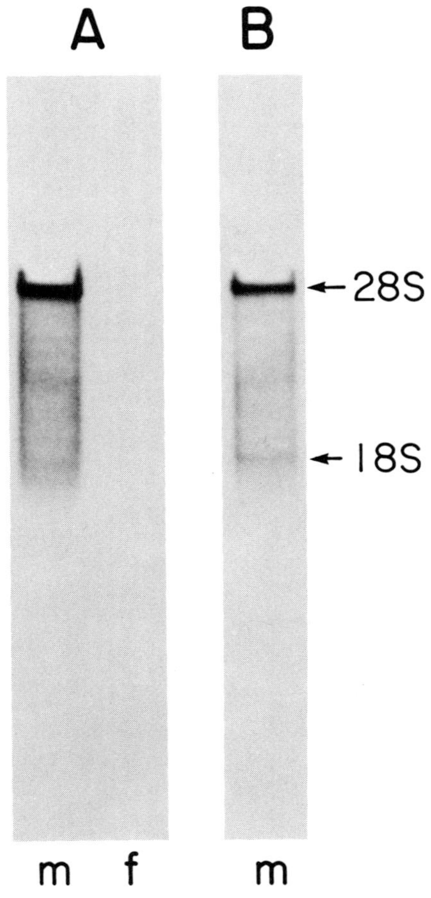

FIGURE 4. Northern blot hybridization analysis of male and female MSG polyA$^+$-RNA. Male and female MSG polyA$^+$-RNA, purified as described in Fig. 2, was separated on a 1.2 percent agarose-formaldehyde gel and transferred to a nitrocellulose filter for Northern analysis as described (Lehrach et al., 1977). Lane m, 2 μg male MSG polyA$^+$-RNA; lane f, 2 μg female MSG polyA$^+$-RNA. Size markers were 28S and 18S rat ribosomal RNAs of 4718 and 1874 nucleotides in length, respectively (I. Wool, personal communication).

clones was determined using the chemical method of Maxam and Gilbert (1977). No sequence differences were detected within the overlapping areas. The contiguous 4706 bp DNA sequence derived from these clones is shown in Fig. 5.

C. EGF is encoded within a putative 133,000 dalton protein precursor

Examination of the 4706 bp nucleotide sequence (Fig. 6) revealed that one translational reading frame could code for an uninterrupted 1217 residue amino acid sequence. The 53 amino acids of mouse EGF were found to be embedded within

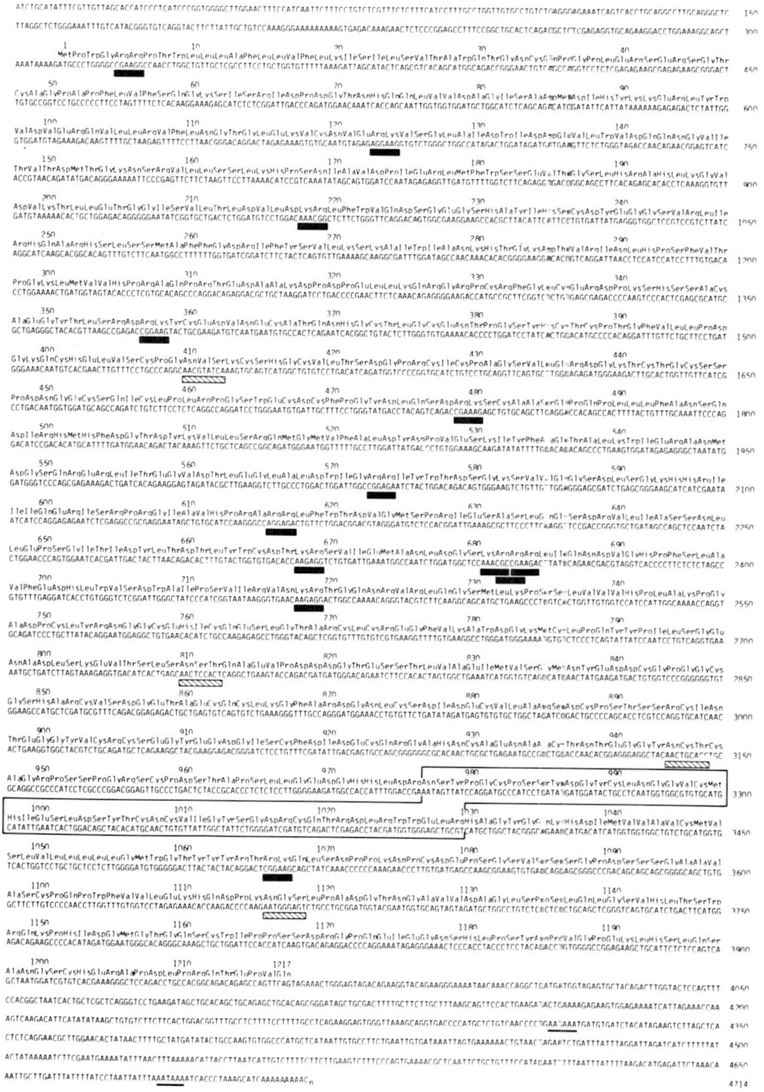

FIGURE 5. Nucleotide sequence of mouse EGF cDNA. The amino acid sequence of the longest, uninterrupted reading frame is numbered from the putative initiation methionine. Nucleotides are numbered at right; the EGF coding sequence is boxed; potential sites for asparagine-linked glycosylation are underlined with a hatched bar. All pairs of basic residues with the potential to be cleaved by a putative precursor processing enzyme have been underscored with a solid bar. Potential polyA addition signals are underlined.

this 1217 amino acid long polypeptide sequence; the 3504 nucleotides coding for the presumptive EGF precursor protein are flanked by an 744 nucleotide long 3' untranslated region (including the termination codon) and by an unusually high number of untranslated nucleotides at the 5' end (309). The 3'-untranslated region contains two AATAAA sequences, thought to be important for polyadenylation of mRNAs (Proudfoot and Brownlee, 1976). It is possible that either of these sites may be used for polyadenylation (Tosi et al., 1981). The number of nucleotides missing in our 5' sequence (~30-40) was determined in a primer extension experiment using two synthetic 30 nucleotide-long oligonucleotide primers from two different locations near the 5' end of our sequence (data not shown).

The open reading frame that includes the EGF amino acid sequence starts with an initiation codon ATG, three nucleotides downstream from an in-frame termination codon (TAA). The initiator methionine is followed by a stretch of 29 amino acids which closely resemble a signal peptide. For example, residues 10-19 are extremely hydrophobic, containing only alanine, leucine, valine and phenylalanine residues. The potential signal peptidase cleavage site is unknown since the NH_2-terminal sequence of Pro-EGF has not as yet been determined. However, based on the observation that cleavage usually occurs subsequent to a small amino acid (Blobel et al., 1979), and given the average length of signal peptides (~15-30 residues), amino acids Ile23, Val26, and Trp29 are each possible candidates for the amino terminal residue of Pro-EGF. The sequence of the mature EGF protein begins at position 977 and ends at amino acid 1029. The sequence is in perfect agreement with that reported by Savage et al. (1973). The carboxyl terminal arginine residue of EGF is followed by a 188 amino acid long C-terminal polypeptide extension. The calculated molecular weight for the entire precursor protein is 133,000 of which only about 5 percent (6045 daltons) are accounted for by the known EGF sequence.

The precursor protein contains 5 potential sites for asparagine linked glycosylation at residues 111, 410, 810, 944 and 1116, which are evenly distributed throughout the protein. Two regions, between residues 330-470 and residues 750-1020, are particularly cysteine-rich, and include the mature EGF sequence.

Several other polypeptide hormones are synthesized as large precursor proteins, including preproenkephalin A (Noda et al., 1982) and B (Kakidani et al., 1982), pro-opiomelanocortin (Nakanishi et al., 1979) and procalcitonin (Amara et al., 1982). In each of these cases, active polypeptide

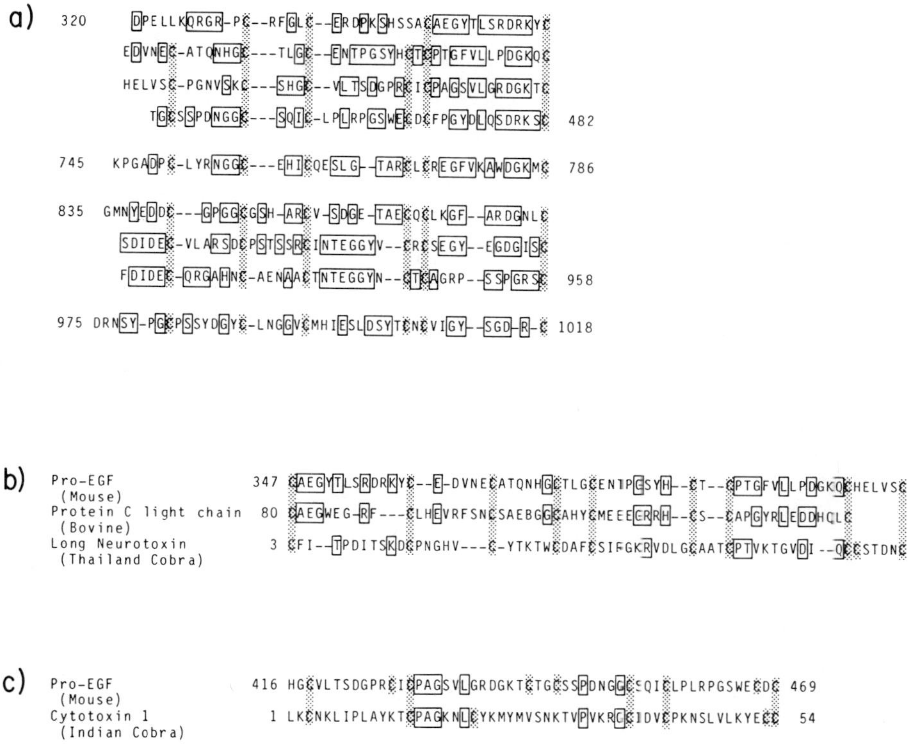

FIGURE 6. Repeated sequence homologies within the EGF precursor and their relationship to sequences of other proteins. Amino acid positions are shown at the beginning and end of the partial sequences. Dashes demarcate gaps introduced to achieve optimal alignment of cysteine residues. Shading indicates an amino acid which occurs at the same position in at least one other repeat unit. (a) comparison of sequences within the EGF precursor; (b) comparison of an EGF precursor region with a portion of protein C (Fernlund and Stenflo, 1982) and a long neurotoxin (Dayhoff, 1978); (c) comparison of another EGF precursor region with a cytotoxin sequence (Dayhoff, 1978).

sequences are usually bounded by the pairs of basic amino acid residues Lys-Arg, Lys-Lys or Arg-Arg; it is at these sites that proteolytic processing occurs. The putative EGF precursor may, however, be processed by a different mechanism. While pairs of these basic residues can be detected in the putative precursor sequence, they do not flank the mature EGF protein. In contrast to the examples mentioned above, processing by the EGF binding protein appears to occur between basic dipeptides Arg-Asn at the amino-terminus and Arg-His at the carboxyl-terminus. In each case one basic amino acid remains part of the EGF polypeptide. This may represent an EGF specific variation of a more general processing scheme employed by a variety of arginine esteropeptidases, while the EGF binding protein (Frey et al., 1979) may have broader site specificity than other processing enzymes. It is also possible that the EGF precursor is substrate for several specific protease activities in addition to the EGF binding protein.

Frey and coworkers (Frey et al., 1979) reported the existence of a 9000 dalton biosynthetic precursor of EGF; these workers provided evidence which suggested that the precursor contained a carboxyl terminal polypeptide extension. The 9000 dalton polypeptide could be an intermediate cleavage product, terminating perhaps at one of several basic amino acid residues located between amino acids 1060 and 1080 of the EGF precursor (Fig. 6).

D. Are there functional polypeptides encoded within the EGF precursor mRNA?

It is difficult to imagine that such a large precursor polypeptide would code for just a small polypeptide hormone, and the presence of multiple basic dipeptides in the sequence suggests that more active polypeptides may also be encoded therein. Analysis of the cysteine-rich areas described above revealed an interesting regular spacing pattern of the cysteine residues, common to both of the cysteine-rich regions. The boundaries of the sequence repeat units are not readily apparent, however the sequence could be divided into 9 groups, 8 of which displayed homologous cysteine residue positions (fig. 6a).

The reason for aligning the sequences in this way was that the cysteine-rich area closest to the carboxyl end, between amino acid 320 and 482, ends with a Cys-12-Cys sequence (numbers represent the length of amino acid sequences separating the cysteine residues). The region between amino acids 751 and 786 appears tc be an independent repeat unit which is interspersed between the two larger cysteine-rich areas, and also ends with a Cys-12-Cys group. Therefore we placed the Cys-12-Cys type repeat unit at the right end of our sequence alignment.

Some sequence gaps had to be introduced to achieve optimal alignment, however the high extent of similarity between the nine sequence units which include part of the mature EGF sequence (bottom line, 6a) is clearly evident in fig. 6a. Homologies within the spacer sequences are most extensive in the Cys-12-Cys or Cys-10-Cys regions (EGF has a Cys-8-Cys at this position).

On the basis of this comparison, the interspersed cysteine-rich sequence 751 to 786 appears to be more closely related to the first group of four repeats (fig. 6a, upper). In this group of repeats, residues Cys-X-X-Gly-X-X-X-X-X-X-X-Lys-X-Cys are conserved. Other regions of striking homology are found at positions 909-913 and 1012-1016 (Gly-Tyr-X-Gly-Asp), at 897-902 and 938-943 (Asn-Thr-Glu-Gly-Gly-Tyr) and at 877-880 and 919-922 (Asp-Ile-Asp-Glu). We do not know whether these regular sequential patterns are of structural necessity to maintain a conformation essential for the correct processing or intracellular traffic of the EGF precursor. Alternatively, these regular cysteine arrangements (or portions thereof), with variable but related spacer sequences, could reflect independent polypeptides with different functions within the EGF-precursor.

Comparison of subsets of these sequences with protein sequence data compiled by Dayhoff (1978) detected two groups of proteins displaying sequence homologies: serine proteases and snake toxins (fig. 6b and c). The serine proteases showed sequence homologies within structures characteristic of this group of enzymes (Kringle structures (Magnusson et al., 1976)). The serine protease partial sequence with the highest degree of homology to pro-EGF 347-407 is shown in fig. 6b. Interestingly, this same region displays remarkable sequence similarity to snake neurotoxins (Dayhoff, 1978), although the homology involves primarily the Cys-Cys spacing. Considering the evolutionary distance between mouse and snake we still believe this to be significant, in particular because the pro-EGF sequence directly following (416-469) shows about equal homology to another group of snake toxins, the cytotoxins (Dayhoff, 1978) (fig. 7c). This finding gains even more significance by the fact that the venom gland is the snake analog to the MSG. The expression of the precursor protein from our cloned sequences and the generation of antibodies to particular regions of the precursor will be useful in elucidating the processing and potential additional functions of other portions of the precursor molecule.

ACKNOWLEDGMENTS

We are grateful to Mark Vasser for carrying out oligonucleotide synthesis, and to Dr. Peter Seeburg, Joel Hayflick and Dr. Ellson Chen for their M13 sequencing efforts which aided in the initial identification of the EGF cDNA clone. We thank Lisa Coussens and Cara Berman for their laboratory support, Jeanne Arch for her expert preparation of this manuscript, Dr. Suzanne Pfeffer for useful discussions, and Dr. Stacey Sias for her contributions during early phases of this work.

We gratefully acknowledge Dr. M. Waterfield for pointing out the cysteine repeat units present in our sequence. We thank Mr. Robert Swanson, President of Genentech, Inc. for his longstanding enthusiasm and support.

REFERENCES

Amara, S.G., Jonmas, V., Rosenfeld, M.G., Ong, E.S., and Evans, R.M. (1982) Nature 298, 240–244.

Blobel, G., Walter, P., Chang, C.N., Goldman, B.M., Erickson, A.H., and Lingappa, V.R. (1979) Society for Exp. Biol. Sym. XXXIII, Secretory Mechanisms (Hopkins, C.R. and Duncan, C.J., eds.) Cambridge University Press, pp. 9–36.

Carpenter, G. and Cohen, S. (1979) Ann. Rev. Biochem. 48, 193–216.

Carpenter, G. and Cohen, S. (1978) In Biochemical Actions of Hormones. G. Litwak, ed., Vol. 5, New York: Academic. pp. 203–47.

Cohen, S. (1959) J. Biol. Chem. 234, 1129–37.

Cohen, S. (1962) J. Biol. Chem. 237, 1555–62.

Crea, R. and Horn, T. (1980) Nucl. Acids Res. 8, 2331-2348.

Dayhoff, M.O. (1978) Atlas of Protein Sequence and Structure, 5, 165–195.

Fernlund, P. and Stenflo, J. (1982) J. Biol. Chem. 257, 12170–12179.

Frey, P., Forand, R., Maciag, T., and Shooter, E.M. (1979) Proc. Natl. Acad. Sci. USA 76, 6294–6298.

Gray, A., Dull, T.J. and Ullrich, A. (1983) Nature 303, 722–725.

Gregory, H. (1975) Nature 257, 325–27.

Grunstein, M. and Hogness, D.S. (1975) Proc. Natl. Acad. Sci. USA 72, 3961-3964.

Kakidani, H., Furutani, Y., Takahashi, H., Noda, M., Morimoto, Y., Hirose, T., Asai, M., Inayama, S., Nakanishi, S., and Numa, S. (1982) Nature 298, 245-249.
Lehrach, H., Diamond, D., Wozney, J.M., and Boedtker, H. (1977) Biochem. 16, 4743-4751.
Magnusson, S., Sottrup-Jensen, L., Petersen, T.E., Dudek-Wojciechowska, G., and Claeys, H. in Proteolysis and Physiological Regulation (eds. Ribbons, D.W. and Brew, K.) 203-212 (Academic, New York, 1976).
Maniatis, T., Frisch, E.F. and Sambrook, J. 1982. Cold Spring Harbor Laboratory.
Maxam, A.M. and Gilbert, W. (1977) Proc. Nat. Acad. Sci. USA 74, 560-564.
Nakanishi, S., Inoue, A., Kita, T., Nakamura, M., Chang, A.C.Y., Cohen, S.N., and Numa, S. (1979) Nature 278, 423-427.
Noda, M., Furutani, Y., Takahashi, H., Toyosato, M., Hirose, T., Inayama, S., Nakanishi, S., and Numa, S. (1982) Nature 295, 202-206.
Proudfoot, N.J. and Brownlee, G.G. (1976) Nature 263, 211-214.
Sanger, F., Nicklen, S., and Coulson, A.R. (1977) Proc. Natl. Acad. Sci. USA 74, 5463-5467.
Savage, C.R., Jr., Hash, J.H., and Cohen, S. (1973) J. Biol. Chem. 248, 7669-7672.
Savage, C.R., Jr., Inagami, T., and Cohen S. (1972) J. Biol. Chem. 247, 7612-21.
Southern, E. 1975. J. Mol. Biol. 98, 503-517.
Taylor, J.M., Cohen, S., and Mitchell, W.M. (1970) Proc. Natl. Acad. Sci. USA 67, 164-71.
Taylor, J.M., Illmensee, R., and Summers, J. (1976) Biochim. Biophys. Acta 4, 324-330.
Taylor, J.M., Mitchell, W.M., and Cohen, S. (1972) J. Biol. Chem. 247, 5928-34.
Taylor, J.M., Mitchell, W.M., and Cohen, S. (1974a) J. Biol. Chem. 249, 3198-3.
Taylor, J.M., Mitchell, W.M., and Cohen, S. (1974b) J. Biol. Chem. 249, 2188-2194.
Tosi, M., Young, R.A., Hagenbuchle, O., and Schibler, U. (1981) Nucleic Acids Res. 9, 2313-2323.
Ullrich, A., Shine, J., Chirgwin, J., Pictet, R., Tischer, E., Rutter, W.J., and Goodman, H.M. (1977) Science 196, 1313-1317.
Wallace, B., Johnson, M.J., Hirose, T., Miyake, T., Kawashima, E.H., and Itakura, K. (1981) Nucleic Acids Research 9, 879-894.
Wickens, M.P., Buell, G.N., and Schimke, R.T. (1978) J. Biol. Chem. 253, 2483-2495.

STRUCTURE OF PRECURSORS DERIVED FROM THE SEQUENCES OF CLONED cDNAs

William J. Rutter
James Scott
Mark Selby
Robert J. Crawford
Lu-ping Shen
Peter Hobart
Ray Sanchez-Pescador
Graeme I. Bell

Department of Biochemistry & Biophysics
University of California
San Francisco, California

Hormone Research Laboratory
University of California
San Francisco, California

For the development of the intricate wiring and various discriminatory functions of the nervous system, complex strategies for regulation of nerve cell growth and function must be involved. These may include (1) general regulators that to some degree balance growth of functionally interrelated cells in the brain and in turn relate them to the general needs of the periphery; (2) trophic factors which selectively direct wiring according to preprogrammed functions or of unusual demands; (3) selective signals operating between specific nerve cell elements in order to filter out the "noise" that would be produced from a common signal in order to accommodate another less frequently anticipated function; and (4) direct communication with the fundamental physiological systems, e.g. metabolism, sexuality, etc. via medium and long term signals. Some of these are perhaps more appropriately mediated by molecules rather than by nerve-cell communications.

BIOCHEMICAL AND CLINICAL ASPECTS OF NEUROPEPTIDES: SYNTHESIS, PROCESSING, AND GENE STRUCTURE

ISBN 0-12-417320-9

Many of the above requirements are met by peptides. There has been great progress in the isolation, characterization and elucidation of the mechanism of action of various physiologically active peptides. This progress has been made in the face of enormous technical problems: 1) Neurotransmitters and other regulatory hormones are usually present in vanishingly small quantities. They may be produced in a small group of cells and are effective in the nanomolar to picomolar range. 2) Some of the specific regulators act in concert, and perhaps may be produced by processing of a common precursor polyprotein, e.g. proopiomelanocortin, preproenkephalin, etc. The isolation and study of such protein precursors is still much more difficult than the isolation of the independent entities. Further, different forms of the peptide may exert different physiological effects; this enhances the need to study them. 3) Neurological function may be mediated by members of gene families that do not necessarily have identical structures. Thus indirect and static assays such as those commonly employing antibodies may not detect the specific molecule involved or indeed the cell of origin, or action, of that molecule.

For all these reasons molecular genetic techniques operating at the DNA/RNA level offer a powerful supplementary approach to the identification and study of neuropeptides: 1) The methods are sensitive; for example, cDNA libraries can be produced from nanograms of mRNA. 2) The isolation of genes and gene sequences provides useful information concerning the precursors that are most difficult to obtain by isolation procedures. The structure of the initial gene products frequently predicts new molecules or unanticipated regulative relationships between multiple products of a common precursor. 3) The isolated genetic sequences can be employed in the biological synthesis of the peptide of interest in heterologous systems. 4) Pseudogenetics can be performed in which biological responses between different cell types can be produced and identified with particular structures.

This paper compares the genetic structures of hormones and regulatory factors which are relevant to brain function. Only one of these, the nerve growth factor, has an action presumably restricted to neuronal cells. Others may fulfill brain-specific functions including those of neurotransmitters but also have other physiological functions. Indeed one of the major paradoxes concerns the apparent multiple roles of many neuropeptides, or stated conversely, the neurological role of many metabolically active peptide hormones.

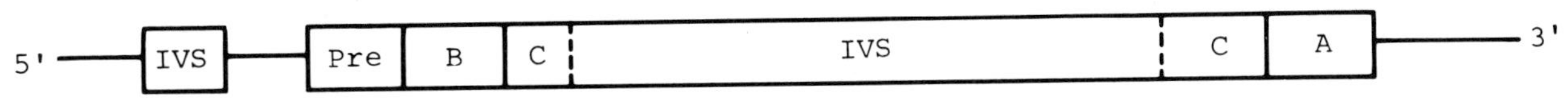

FIGURE 1. Schematic representation of the human insulin gene including the locations of both intervening sequences and the protein coding regions. pre=signal peptide; B=B chain; C=connecting peptide; A=A chain.

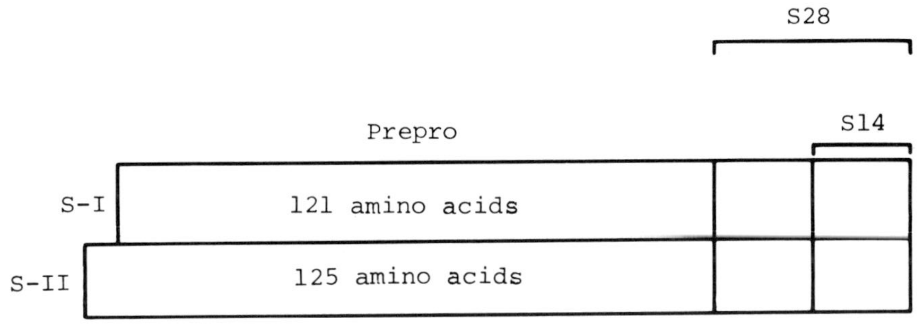

FIGURE 2. A schematic representation of the cDNAs coding for the precursors of somatostatins I and II.

INSULIN: THE PARADIGM

Insulin was the first hormone whose precursor, previously detected by immunological and other methods (1) was fully elucidated via sequence analysis of the mRNA (2,3) (Figure 1). These studies determined the structure of the signal peptide and confirmed the sequence of proinsulin, the immediate precursor. The molecular structure illuminated the role of the signal peptide in secretion and defined the function of processing as the removal of a peptide necessary for folding but undesirable for hormonal function. Preproinsulin thus has been frequently taken as the model of polypeptide hormone precursors. In some respects, the simplicity of the function represented in this structure is unusual. Virtually all of the other peptide precursors are more complicated structures. Many are apparently polyproteins and appear to be multifunctional. Like the others, however, insulin or sequences related to it may be synthesized in brain tissue.

SOMATOSTATIN: THE MULTIFUNCTIONAL PARADOX

The 14 amino acid peptide somatostatin inhibits the secretion of a number of hormones and enzymes and may act as a neurotransmitter. Its structure is highly conserved in nature, the identical sequence present in humans is found in fish (4,5) (Figure 2). The multiple functions of somatostatin are difficult to correlate with a single structure but since the cells that synthesize somatostatin are dispersed, it has been postulated that this hormone functions in a paracrine fashion. Thus somatostatin producing cells regulate the production of the hormone from target cells in close proximity. This postulate requires that the somatostatin producing cells have independently regulated receptors such that each minisystem can operate independently. Whereas this mechanism may in part be operative, it cannot explain the selective regulation of insulin and glucagon secretion by somatostatin, because cells secreting insulin and glucagon may be virtually contiguous in the islets of Langerhans, both juxtaposed to somatostatin producing cells. Because the metabolic functions of insulin and glucagon tend to counterbalance each other, it seems unlikely that the hormones are coordinately regulated by the same structure. My colleagues and I therefore began to investigate the possibility of multiple forms of this hormone. We employed anglerfish islets for these studies because 1) fish are known to secrete higher levels of somatostatin and glucagon than mammals and 2) anglerfish islets are readily isolated and resolved from acinar tissue which contains troublesome nucleases. From the cDNA library prepared from the polyA+ RNA obtained frcm these islets, we were able to isolate two distinct nonhybridizing clones that coded for somatostatin-like sequences. One of these cDNAs reflected a mRNA coding for a peptide identical to that found in humans and other animals. We termed this somatostatin I (6). The somatostatin I cDNA was later isolated from a human somatostatinoma (7). The second, called somatostatin II (6), coded for a somatostatin analog which differed in two amino acid positions from the classical somatostatin; phenylalanine 7 is replaced by a tyrosine and threonine 10 is replaced by a glycine residue. In both somatostatin cDNAs, the somatotatin moieties were at the C terminus of a 121 and 125 amino acid peptide, respectively. The two precursors are obviously related in structure but are quite different in specific amino acid sequence.

Synthesis of the two oligopeptides and subsequent testing of their potency against insulin and glucagon release showed that somatostatin I blocks glucagon secretion very well and insulin to a lesser degree. In contrast somatostatin II blocked insulin release efficiently and had no effect on glucagon secretion. Thus it appears that the two somatostatins have selective roles in regulating the release of insulin and glucagon, respectively. There is however no dramatic difference in their ability to depress growth hormone secretion. These findings suggest that there may be additional somatostatin entities coded by distinct genes that relate to the other specific functions of this hormone.

Both precursors predict that a 28 amino acid somatostatin containing peptide can be formed by cleavage of a single basic residue. Fourteen and 28 amino acid forms of somatostatin I are both biologically active, but the two molecules have differing effects on cellular receptors and may have different relative inhibitory activity against insulin and glucagon secretion. Whether these effects are physiologically relevant, they focus attention on the approximate 80 amino acid sequence between the somatostatin moiety and the rather typical signal peptide region. This region has no dibasic residues and no obvious cleavage sites. It could simply be a "connecting peptide" that joins two significant functions. However, there is impressive conservation of structure within this region between the human and anglerfish mRNA (cDNA). In addition, very few changes occur in the entire molecule between the human (7) and the rat (5). This contrasts with the variation in the structure of the insulin connecting peptide and suggests that this region of the molecule may have a biological function with, or independent from, the somatostatin moiety. It should be possible to synthesize the precursor in a heterologous cell system using the coding regions of anglerfish and human cDNAs in order to test this hypothesis.

The genes of somatostatin I from anglerfish and human genomic DNA have been isolated by molecular cloning and have been sequenced (4,8). The studies show a single intervening sequence positioned at the 56-57 residue (Gln/Glu/junction) upstream from the somatostatin moiety. Interestingly, the intervening sequence is positioned near a site of length polymorphism and amino acid divergence. As postulated elsewhere (9), intron junctions may be sites of additions or deletions of amino acids in evolution. This may result from junction sliding.

The 5' flanking sequences of the anglerfish gene reveal a number of striking repeats that extend into the transcribed portion of the molecule. These repeats form a number of hairpin structures which can variously include or exclude the promoter sequence (4). Thus it is possible that regulation of expression involves these structures. Interestingly, the human gene contains none of these repeats but the hairpin structures are present (though they are less dramatic).

GLUCAGON AND RELATED PEPTIDES: A HORMONE FAMILY PRODUCED FROM A SINGLE GENE

The 29 amino acid polypeptide glucagon is synthesized in the A cells of the endocrine pancreas. However, glucagon-containing peptides and immunologically related peptides of various sizes have been detected in the intestine, stomach, salivary gland and brain. Glucagon is well known for its stimulation of glycogenolysis and gluconeogenesis and ketogenesis but it is also recognized as an inducer of certain enzymes in the liver. Its role in the brain remains unclarified. Bell and colleagues have recently isolated cDNAs coding for glucagon from the human (10) and hamster (11). Lund et al. (12) have reported the structure of the glucagon cDNA from anglerfish (Figure 3). The nucleotide sequence of the cDNA predicts the primary gene product to be 180 amino acids (hamster) and 179 amino acids. The molecular size is in acceptable agreement with the size of the precursor predicted from previous biosynthetic studies (13). This precursor contains the sequence of glucagon and two other glucagon-like polypeptides of 37 and 35 amino acids, respectively. GLP-1 and -2 are related but are not identical with other members of the glucagon-secretin family of gastrointestinal

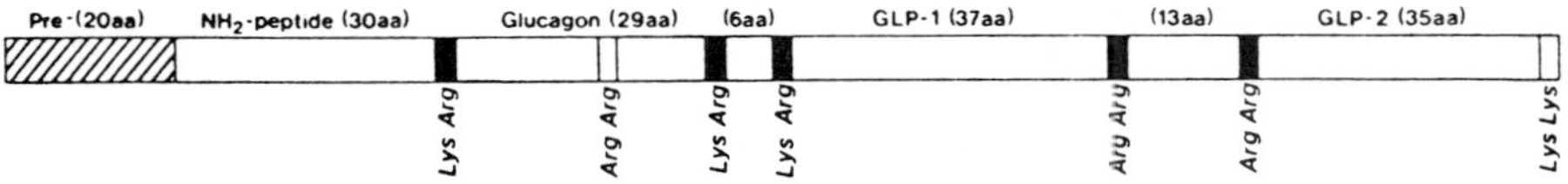

FIGURE 3. Schematic representation of the structure of the hamster pancreatic preproglucagon and the two glucagon-like peptides (GLP). The dibasic peptides are boxed and those which are potentially cleaved are darkened.

hormones. In addition, the precursor contains the sequences of several other glucagon-containing polypeptides that are ordinarily not found in the pancreas. These include the intestinal glucagon-like immunoreactant I (GLI-1) or glycentin (proglucagon 1-69) and a glycentin-related pancreatic polypeptide GRPP (proglucagon 1-30). The precursor also contains a higher molecular weight glycentin-like compound probably containing the elements of glucagon and GLP-1. Both the glycentin-like peptides are found in the intestine. It thus appears that both the pancreatic and intestinal glucagon-related polypeptides are derived from a common precursor by different processing steps. The precursor can generate at least 11 unique polypeptides from known dibasic cleavage sites. This hypothesis of cell-specific processing requires that cells in the intestine either have a distinct set of enzymes that cleave at dibasic residues but also have additional specificity determinants or, alternatively, that due to the structure of the precursor, the efficiency of cleavage at various sites differs dramatically and that the cells simply contain different levels of a common cleavage enzyme. Eight of the possible cleavage products have been identified biochemically and immunochemically. The biological roles of the intestinal glucagon-containing polypeptides is unclear, although glycentin may inhibit gastric acid secretion. The GLP-1 and GLP-2 moieties have previously been undetected, hence the processing steps involved in their formation, and their biological role remain unknown. However, the sequence conservation of the GLP-1 region is consistent with a biological function of this peptide. The knowledge of their structure will now allow their synthesis and then the function of this family of glucagon-containing polypeptides can be critically examined.

The structure of the anglerfish pancreatic preproglucagon has been derived from the nucleotide sequence of the cDNA by Lund et al. (12). The anglerfish contains an additional **preproglucagon structure** (preproglucagon II), but its sequence has not yet been reported (12). Remarkably, anglerfish preproglucagon I is only 124 amino acids. This difference is due to the absence of the 13 amino acid spacer peptide and the second glucagon-like peptide, GLP-2. Thus the organization of the prohormone has not been conserved during vertebrate evolution. The presence of the additional glucagon-like peptide in the mammalian gene has been observed. The evolutionary history of the hormone may be more apparent when the structure of the gene is defined.

EGF: A POLYPROTEIN PRECURSOR CONTAINS EGF AND SEVEN RELATED PROTEINS

Epidermal Growth Factor is a 53 amino acid protein that stimulates the proliferation and differentiation of cells of ectodermal and mesodermal origin. In addition, it is presumably identical to urogastrone, a hormone that inhibits HCl release from the gastric mucosa (14). EGF also exerts a number of effects in developing organs including precocious eyelid opening, incisor eruption, and accelerated maturation of the lung and trachea (15). The presence of high levels of EGF in mother's milk may also suggest a role of this hormone in the development of the gastrointestinal tract. Significantly, EGF receptors are present in various adult tissues including brain. This hormone thus is presumably involved in the control of growth and function of a variety of cells.

It has been recently reported that EGF stimulates phosphorylation of its own receptor by a tyrosine-specific protein kinase normally associated with the receptor (16). This tyrosine-specific protein kinase may be related to those encoded by the transforming genes of some retroviruses (17). It is possible that the control of cell proliferation by EGF is related to the uncontrolled proliferation induced by the retroviruses.

EGF is found in low concentrations in several tissues including salivary gland and the intestine. It is found inexplicably in high concentrations in the submaxillary gland of the male mouse (interestingly not of the submaxillary gland of other species). It has also been isolated from human urine (urogastrone).

EGF specific cDNA (18) was isolated from a male mouse submaxillary cDNA library by hybridization with synthetic oligodeoxynucleotide probes. In line with the known increased abundance in the male gland, hybridization experiments with a labeled EGF probe showed that the mRNA encoding EGF is

1 (357-399)	Arg Lys Tyr Cys Glu Asp Val Asn Glu Cys Ala Thr Gln Asn His Gly Cys Thr
2 (400-440)	Gln Cys His Glu Leu Val Ser Cys Pro Gly Asn Val Ser Lys Cys Ser
3 (441-480)	Thr Cys Thr Gly Cys Ser Ser Pro Asp Asn Gly Gly Cys Ser Gln
4 (745-784)	Lys Pro Gly Ala Asp Pro Cys Leu Tyr Arg Asn Gly Gly Cys Glu
5 (803-885)	Met Val Ser Gly Met Asn Tyr Glu Asp Asp Cys Gly Pro Gly Gly Cys Gly Ser His
6 (886-925)	Ser Asp Cys Pro Ser
7 (926-976)	Gly Ala His Asn Cys Ala Glu Asn
EGF (977-1029)	Asn Ser Tyr Pro Gly Cys Pro Ser Ser Tyr Asp Gly Tyr Cys Leu Asn Gly Gly Val

Leu Gly Cys Glu Asn Thr Pro Gly Ser Tyr His Cys Thr Cys Pro Thr Gly Phe Val Leu

His Gly Cys Val Leu Thr Ser Asp Gly Pro Arg Cys Ile Cys Pro Ala Gly Ser Val Leu

Ile Cys Leu Pro Leu Arg Pro Gly Ser Trp Glu Cys Asp Cys Phe Pro Gly Tyr Asp Leu

His Ile Cys Gln Glu Ser Leu Gly Thr Ala Arg Cys Leu Cys Arg Glu Gly Phe Val Lys

Ala Arg Cys Val Ser Asp Gly Glu Thr Ala Glu Cys Gln Cys Leu Lys Gly Phe Ala Arg

Ser Arg Cys Ile Asn Thr Glu Gly Gly Tyr Val Cys Arg Cys Ser Glu Gly Tyr Glu Gly

Ala Ala Cys Thr Asn Thr Glu Gly Gly Tyr Asn Cys Thr Cys Ala Gly Arg Pro Ser Ser

Cys Met His Ile Glu Ser Leu Asp Ser Tyr Thr Cys Asn Cys Val Ile Gly Tyr Ser Gly

Leu Pro Asp Gly Lys

Gly Arg Asp Gly Lys

Gln Ser Asp Arg Lys

Ala Trp Asp Gly Lys

Asp Gly Asn Leu Cys Ser Asp Ile Asp Glu Cys Val Leu Ala Arg

Asp Gly Ile Ser Cys Phe Asp Ile Asp Glu Cys Gln Arg

Pro Gly Arg Ser Cys Pro Asp Ser Thr Ala Pro Ser Leu Leu Gly Glu Asp Gly His His Leu Asp Arg

Asp Arg Cys Gln Thr Arg Asp Leu Arg Trp Trp Glu Leu Arg

FIGURE 4. Comparison of EGF-like peptides 1 to 7 and EGF. The proteins were aligned at the common Cys-X-Cys sequence. Identical amino acids shared by at least four members of the family and Cys residues are boxed. The boundaries of the proteins are indicated in parentheses. Only the sequence of EGF-like peptide 5 from amino acids 832 to 885 is indicated. EGF-like peptide 1 is the top line in each group.

at least 10 times more abundant in the male gland than in the female gland. Remarkably, EGF cDNA is approximately 4800 bases long. It contains a single open reading frame of 3396 bases encoding a protein of 1217 amino acids (MW 133,000) (Figure 4). The amino terminal region of the precursor contains a hydrophobic area typical of signal peptides. The remaining proEGF contains 11 pairs of basic amino acids, putative sites for proteolytic processing in other systems. However, these do not obviously coincide with the processing pathway for EGF or many of the other precursors. The EGF moiety is located in the central region of the precursor, and is flanked by polypeptide segments of 976 and 188 amino acids.

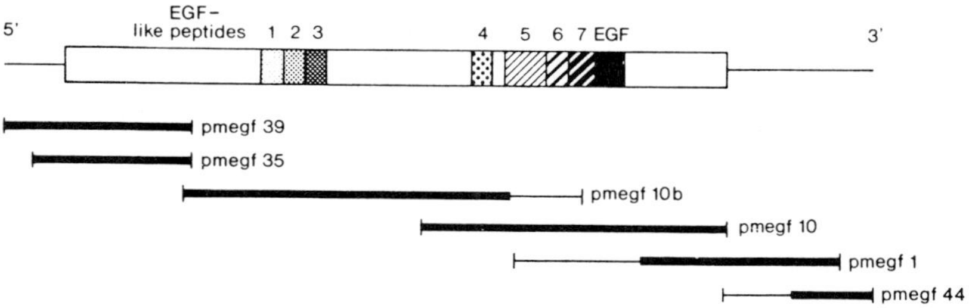

FIGURE 5. Schematic representation of mouse submaxillary EGF mRNA and protein. The box indicates the protein-coding portion of the mRNA. The positions of the 7 EGF-like peptides and EGF are indicated. The thin line corresponds to the untranslated regions of the mRNA.

The sequence contains an arginine adjacent to the aminoterminal asparagine of EGF and a carboxyterminal arginine. Thus an arginine-specific peptidase may be involved in processing of the precursor. It is already well known that EGF can be isolated in a form associated with an arginine-specific peptidase, as well as an EGF binding protein. Significantly, neither of these molecules is encoded by the large EGF mRNA. Examination of the precursor revealed the presence of 7 EGF related peptides of different sizes, all containing the intriguing cys-X-cys structure present in EGF itself (Figure 5). The boundaries of each of the EGF-like peptides are defined by a basic amino acid. However sometimes lysine instead of arginine residues are involved; hence, the arginine specific gastropeptidase can not act to produce all these structures. The sizes of the peptides vary from 39 to 83 amino acids with 5 containing about 40 residues. The EGF-like sequences are arranged in two clusters of three members each; EGF is at the carboxyterminus of the second cluster. The remaining EGF-like protein is located between the two tandem arrays separated by a spacer of about 20 amino acids from the group containing EGF. ProEGF therefore could be processed in a variety of different ways to yield various combinations of EGF and EGF-like molecules. Further EGF and EGF-like peptides account for only about one-third of the size of the precursor, so additional active peptides could also be formed from pro-EGF. In particular the elements demarcated by the dibasic residues deserve attention: 11 additional peptides can be produced by cleavage at the basic residues. The availability of the structure now allows synthesis and characterization of the properties of various possible peptides derived from the

```
GAGCGCCUGGAGCCGGAGGGGAGCGCAUCGAGUGACUUUGGAGCUGGCCUUAUAUUUGGAU

                                  1
                                  met leu cys leu lys pro
CUCCCGGGCAGCUUUUGGAAACUCCUAGUGAAC AUG CUG UGC CUC AAG CCA

            10                                      20
val lys leu gly ser leu glu val gly his gly gln his gly gly
GUG AAA UUA GGC UCC CUG GAG GUG GGA CAC GGG CAG CAU GGU GGA

                                30
val leu ala cys gly arg ala val gln gly ala gly trp his ala
GUU UUG GCC UGU GGU CGU GCA GUC CAG GGG GCU GGA UGG CAU GCU

            40                                      50
gly pro lys leu thr ser val ser gly pro asn lys gly phe ala
GGA CCC AAG CUC ACC UCA GUG UCU GGG CCC AAU AAA GGU UUU GCC

                                60
lys asp ala ala phe tyr thr gly arg ser glu val his ser val
AAG GAC GCA GCU UUC UAU ACU GGC CGC AGU GAG GUG CAU AGC GUA
                                         ▲
            70                                      80
met ser met leu phe tyr thr leu ile thr ala phe leu ile gly
AUG UCC AUG UUG UUC UAC ACU CUG AUC ACU GCG UUU UUG AUC GGC

                                90
val gln ala glu pro tyr thr asp ser asn val pro glu gly asp
GUA CAG GCA GAA CCG UAC ACA GAU AGC AAU GUC CCA GAA GGA GAC

            100                                     110
ser val pro glu ala his trp thr lys leu gln his ser leu asp
UCU GUC CCU GAA GCC CAC UGG ACU AAA CUU CAG CAU UCC CUU GAC

                                120
thr ala leu arg arg ala arg ser ala pro thr ala pro ile ala
ACA GCC CUC CGC AGA GCC CGC AGU GCC CCU ACU GCA CCA AUA GCU

            130                                     140
ala arg val thr gly gln thr arg asn ile thr val asp pro arg
GCC CGA GUG ACA GGG CAG ACC CGC AAC AUC ACU GUA GAC CCC AGA

                                150
leu phe lys lys arg arg leu his ser pro arg val leu phe ser
CUG UUU AAG AAA CGG AGA CUC CAC UCA CCC CGU GUG CUG UUC AGC

        160                                         170
thr gln pro pro pro thr ser ser asp thr leu asp leu asp phe
ACC CAG CCU CCA CCC ACC UCU UCA GAC ACU CUG GAU CUA GAC UUC

                                180
gln ala his gly thr ile pro phe asn arg thr his arg ser lys
CAG GCC CAU GGU ACA AUC CCU UUC AAC AGG ACU CAC CGG AGC AAG

        190                                         200
arg ser ser thr his pro val phe his met gly glu phe ser val
CGC UCA UCC ACC CAC CCA GUC UUC CAC AUG GGG GAG UUC UCA GUG

                                210
cys asp ser val ser val trp val gly asp lys thr thr ala thr
UGU GAC AGU GUC AGU GUG UGG GUU GGA GAU AAG ACC ACA GCC ACA

            220                                     230
asp ile lys gly lys glu val thr val leu ala glu val asn ile
GAC AUC AAG GGC AAG GAG GUG ACA GUG CUG GCC GAG GUG AAC AUU

                                240
asn asn ser val phe arg gln tyr phe phe glu thr lys cys arg
AAC AAC AGU GUA UUC AGA CAG UAC UUU UUU GAG ACC AAG UGC CGA

            250                                     260
ala ser asn pro val glu ser gly cys arg gly ile asp ser lys
GCC UCC AAU CCU GUU GAG AGU GGG UGC CGG GGC AUC GAC UCC AAA

                                270
his trp asn ser tyr cys thr thr thr his thr phe val lys ala
CAC UGG AAC UCA UAC UGC ACC ACG ACU CAC ACC UUC GUC AAG GCG

            280                                     290
leu thr thr asp glu lys gln ala ala trp arg phe ile arg ile
UUG ACA ACA GAU GAG AAG CAG GCU GCC UGG AGG UUC AUC CGG AUA

                                300
asp thr ala cys val cys val leu ser arg lys ala thr arg arg
GAC ACA GCC UGU GUG UGU GUG CUC AGC AGG AAG GCU ACA AGA AGA

307
gly OP
AGA GGC UGA  CUUGCCUGCAGCCCCCUUCCCCACCUGCCCCCUCCACACUCUCUUGGG

CCCCUCCCUACCUCAGCCUGUAAAUUAUUUUAAAUUAUAAGGACUGCAUGAUAAUUUAUC

GUUUAUACAAUUUUAAAGACAUUAUUUAUUAAAUUUUCAAAGCAUCCUGAAAAAAAA
```

precursor. In addition the elucidation of the structure of the EGF gene should be extremely interesting, since it may help to illuminate the evolutionary origin of this genetic entity.

NGF: DERIVATION FROM POLYPROTEIN PRECURSOR

Nerve Growth Factor (NGF) enhances the survival and stimulates fiber outgrowth and neurotransmitter biosynthesis in sympathetic and sensory neurons. Further, antibodies against NGF when administered to developing animals result in atrophy of the sympathetic system suggesting that this hormone is required continuously (19). NGF is not normally detected in innervated tissues but it can be detected after ablation of the innervating neurons. Subsequently the NGF levels decrease to undetectable levels when the denervated tissue becomes reinnervated (20). It appears, therefore, that NGF is a neurotrophic messenger produced in non-nervous tissues and further, that the expression of the gene is regulated by innervating neurons. Because of the extraordinarily low levels present, NGF cannot be isolated and studied in such tissues. However, like EGF, this hormone is present in inexplicably high levels in the male mouse submaxillary gland NGF (it is also not present at high levels in the submaxillary glands of other species). The NGF cDNA was isolated from a submaxillary cDNA library by the use of a pool of labeled synthetic oligonucleotide probes (21). The mRNA is approximately 1176 bases long, 1068 bases are derived from a single cDNA clone and the remaining sequence determined by primer extension experiments. This RNA contains a single reading frame predicting a 307 amino acid polypeptide (Figure 6). The size of this precursor is significantly larger than those detected by previous biosynthetic and immunoprecipitation experiments (22). Further the NGF moiety had been previously presumed to be centrally located; in fact, it is positioned at the C terminus of the molecule and is flanked by dibasic residues. At its amino terminus, there is a 187 amino acid peptide that may be cleaved at dibasic residues to form peptides of 38, 27 and ∿90 amino acids (Figure 7). NGF is

---

FIGURE 6. The nucleotide and amino acid sequence of the NGF containing mRNA. Boxed basic residues are the potential recognition sites for trypsin-like processing enzymes. The solid arrow denotes the location of an intervening sequence found in a mouse genomic phage clone.

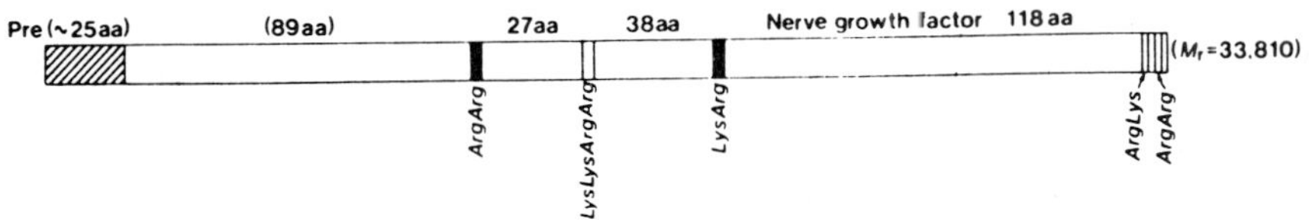

FIGURE 7. Schematic representation of the mouse submaxillary preproNGF polyprotein. The basic di- and tetrapeptide possible cleavage sites are indicated; the sites at which cleavage is known to occur are indicated by solid boxes. The numbers in parentheses are tentative size designations; the others give accurate sizes.

normally found in cells complexed with 26,000 MW alpha and gamma subunits of the 7S complex (23). Neither can be derived from the NGF precursor. This complex could also play a role in secretion.

Unlike other secretory molecules, the NGF precursor does not contain a central strong hydrophobic region typical of signal peptides, it contains two weakly hydrophobic regions. Intriguingly there is a more strongly hydrophobic region in the body of the peptide at residues 77 through 82. Recent biochemical studies appear to confirm the molecular weight of the preproNGF and proNGF as predicted here. However additional studies must be carried out to eliminate the possibility that the initiation of translation occurs at a methionine close to the internal hydrophobic region. This would produce a precursor of 241 amino acids. The amino acid sequence of mature NGF is approximately 25% homologous with proinsulin (24). Alignment of corresponding mRNA segments produced a region of 40% homology; however the overall structure of the cDNAs and positioning of the processing sites as well as the localization of the intervening sequences in the NGF gene are incongruent with that of the insulin gene. Thus, the relationship may be explained as well by convergent evolution as by divergent evolution.

The mouse NGF cDNA was used to screen both mouse and human genomic phage libraries. Six independent clones were isolated from the mouse library. All the clones proved to be identical by restriction mapping and hybridizations. Further mapping and sequencing established that all the NGF encoding DNA except that representing the first 62 amino acids was present in the mouse phage. Moreover the NGF DNA is situated at the 3' end of the insert DNA and is bordered at its 5' end by ~10 kb of mouse DNA. This suggests that the remainder of the NGF encoding DNA is separated from the main body of coding DNA by an enormous intron. The isolation of the phage containing the remainder of the cDNA will allow us to make a more definitive statement regarding the size of the gene and its intervening sequences.

The cDNA was also successfully utilized for screening the Maniatis human library (with Dr. John Darby and Eric Shooter). Three overlapping phage clones were isolated. Mapping with restriction enzymes and hybridization with the cDNA probe suggested that the phage possessed two regions of hybridizing DNA that were separated by several kb of intervening sequence. A synthetic oligonucleotide complementary to the 5' end of the mouse cDNA was hybridized to both mouse and human phage digests. No hybridization was observed with the mouse DNA but hybridization was obvious with the human DNA. Thus the cloned human DNA must possess a greater portion of the human gene. Further characterization of these and other phage will allow us to determine the limits of the gene as well as elucidate potential 5' regulatory elements, and compare the two huge NGF genes.

GENE FAMILIES

The multifarious aspects of regulation by hormones can in part be handled by different structures that interact with distinct receptors to produce unique biological responses in cells and tissues. Members of the hormone family can be encoded in separate gene units and can thus be independently regulated. Or, alternatively, they can be encoded in the same gene and produced by processing of a polyprotein precursor. Thus production of all members of the set are coordinated to the same stimulus. Within the hormone systems described here both genetic constructions are apparent. A number of novel hormone precursors are presented. Their structures are surprisingly complex. The presence of other biologically active peptides may be inferred from 1) the occurrence of dibasic lysine and arginine residues which are typically sites of processing in other hormonal systems. 2) The presence of sequences analogous to other biologically active moieties. 3) A conservation of nucleotide (and amino acid) sequences within the precursors isolated from widely divergent species, for example, human, mouse and fish). It should now be possible to study these biologically by construction of a system for the expression of these genes in heterologous cells that are capable of carrying out the processing reactions. These experiments in themselves would be useful for characterizing the specificity of the processing reactions. In addition, the peptides can be synthesized chemically and their properties examined directly.

THE SALIVARY/GUT/PANCREAS/BRAIN AXIS

One of the most intriguing findings concerning the role of neuropeptides and peptide hormones isolated from the digestive tract (including salivary gland, pancreas) is that most peptides exhibit functions in both systems. Virtually all of the biologically active peptides yet discovered in the salivary/gut/pancreas system are also found in the brain. These include insulin, glucagon, somatostatin, secretin, gastrin, cholecystokinin, pancreatic polypeptide, growth hormone releasing factor (GRF), EGF, NGF and, most intriguingly, Schaller's Hydra neuronal inducer (this symposium). This relationship strongly implies a biological connection between the neurological and digestive systems. This could provide molecular communication between two vital systems in the organism. It could also provide a glimpse of a pathway of evolutionary development, or it could reflect the adaptive reutilization of common mechanisms by the two systems for different purposes. Whatever the mechanism involved, the hormones can now be studied decisively in both systems using the methods of molecular genetics.

REFERENCES

1. Chan, S.J., and Steiner, D.F., *Proc. Nat. Acad. Sci. USA 73*, 1964 (1976).
2. Cordell, B., Bell, B., Tischer, E., DeNoto, F.M., Ullrich, A., Pictet, R., Rutter, W.J., and Goodman, H.M., *Cell 18*, 533 (1979).
3. Bell, G., Pictet, R., Rutter, W.J., Cordell, B., Tischer, E., and Goodman, H.M., *Nature 284*, 26 (1980).
4. Shen, L-P, and Rutter, W.J., *Science*, in press.
5. Taylor, W.L., Collier, K.J., Deschenes, R.J., Weith, H.L. and Dixon, J.C., *Proc. Nat. Acad. Sci. USA 78*, 6694 (1981).
6. Hobart, P., Crawford, R., Shen, L-P., Pictet, R., and Rutter, W.J., *Nature 288*, 137 (1980).
7. Krejs, G.J., Orci, L., Conlon, M., Ravazzola, M., Davis, G.R., Raskin, P., Collins, S.M., McCarthy, D.M., Baetens, D., Rubenstein, A., Aldor, T.A.M., and Unger, R.H., *New Eng. J. Med. 301*, 285 (1979).

8. Crawford, R., Hobart, P., and Rutter, W.J., submitted.
9. Craik, C.S., Rutter, W.J., and Fletterick, R., *Science 220*, 1125 (1983).
10. Bell, G., Sanchez-Pescador, R., Laybourn, P.J., and Najarian, R.C., *Nature 304*, 368 (1983).
11. Bell, G., Santerre, R.F., and Mullenbach, G., *Nature 302*, 716 (1983).
12. Lund, P.K., Goodman, R.H., Dee, P.C., and Habener, J.F., *Proc. Nat. Acad. Sci. USA 79*, 345 (1982).
13. Patzelt, C., Tager, H.S., Carrol, R.S., and Steiner, D.F. *Nature 282*, 260 (1979).
14. Gregory, H., *Nature 257*, 325 (1975).
15. Carpenter, G., and Cohen, S., *Ann. Rev. Biochem. 48*, 193 (1979).
16. Chinkers, M., and Cohen, S., *Nature 290*, 516 (1981).
17. Bishop, J.M., *Ann. Rev. Biochem. 52*, 301 (1983).
18. Scott, J., Urdea, M., Quiroga, M., Sanchez-Pescador, R., Fong, N., Selby, M., Rutter, W.J., and Bell, G.I., *Science 221*, 236 (1983).
19. Levi-Montalcini, R., and Angeletti, R.H., *Pharmacol. Rev. 18*, 619 (1966).
20. Ebendal, T., Olson, L., Seiger, A., and Hedlund, K-O, *Nature 286*, 25 (1980).
21. Scott, J., Selby, M., Urdea, M., Quiroga, M., Bell, G.I., and Rutter, W.J., *Nature 302*, 538 (1983).
22. Burger, E., and Shooter, E.M., *Proc. Nat. Acad. Sci. USA 74*, 3647 (1977).
23. Yanker, B.A., and Shooter, E.M., *Ann. Rev. Biochem. 51*, 845 (1982).
Thoenen, H., *Physiol. Rev. 60*, 1284 (1980).
24. Frazier, W.A., Hogue-Angeletti, R.A., Sherman, R., and Bradshaw, R.A., *Biochemistry 12*, 3281 (1973).

Index